THE PRICE OF TRUTH

PRACTICAL AND PROFESSIONAL ETHICS SERIES

Published in conjunction with the Association for Practical and Professional Ethics

Published in the Series

Practical Ethics
Henry Sidgwick
With an Introduction by Sissela Bok

Thinking Like an Engineer
Michael Davis

Deliberative Politics
Edited by Michael Macedo

From Social Justice to Criminal Justice
Edited by William C. Heffernan and John Kleinig

Conflict of Interest in the Professions
Edited by Michael Davis and Andrew Stark

Meaningful Work
Mike W. Martin

From Morality to Mental Health
Mike W. Martin

The Price of Truth
David Resnik

THE PRICE OF TRUTH

How Money Affects the Norms of Science

DAVID B. RESNIK

OXFORD UNIVERSITY PRESS

2007

OXFORD
UNIVERSITY PRESS

Oxford University Press, Inc., publishes works that further
Oxford University's objective of excellence
in research, scholarship, and education.

Oxford New York
Auckland Cape Town Dar es Salaam Hong Kong Karachi
Kuala Lumpur Madrid Melbourne Mexico City Nairobi
New Delhi Shanghai Taipei Toronto

With offices in
Argentina Austria Brazil Chile Czech Republic France Greece
Guatemala Hungary Italy Japan Poland Portugal Singapore
South Korea Switzerland Thailand Turkey Ukraine Vietnam

Published by Oxford University Press, Inc.
198 Madison Avenue, New York, New York 10016

www.oup.com

Library of Congress Cataloging-in-Publication Data
Resnik, David B.
The price of truth : how money affects the norms of science / David B. Resnik.
p. cm. — (Practical and professional ethics series)
Includes bibliographical references and index.
ISBN-13: 978-0-19-530978-2
ISBN-10: 0-19-530978-2
1. Science—Moral and ethical aspects. 2. Research—Finance. 3. Research—Political aspects. 4. Science and state. 5. Conflict of interests. I. Title. II. Series.
Q175.35.R47 2007
174'.95—dc22 2006040057

1 3 5 7 9 8 6 4 2

Printed in the United States of America
on acid-free paper

This book is dedicated to all the people who have sacrificed their lives in the search for truth.

ACKNOWLEDGMENTS

I wrote the first draft of this book during the 2003–2004 academic year, when I was a professor of medical humanities at the Brody School of Medicine at East Carolina University (ECU). On July 26, 2004, I left my position at ECU and began working as a bioethicist at the National Institute of Environmental Health Sciences (NIEHS), National Institutes of Health (NIH). The ideas and opinions expressed in this book are my own, and do not represent the views of the NIEHS, the NIH, ECU, or the U.S. government. I would like to thank the following people for useful comments and discussions related to this work: John Davis, Ken De Ville, Loretta Kopelman, Sheldon Krimsky, Helen Longino, Michael Resnik, Adil Shamoo, Richard Sharp, and several anonymous reviewers.

CONTENTS

ABBREVIATIONS

AAMC	Association of American Medical Colleges
AAU	Association of American Universities
ACE	Angiotensin-converting-enzyme
ACS	American Chemical Society
ACTR	Australian Clinical Trials Registry
AE	Adverse effect
AIDS	Acquired Immunodeficiency Syndrome
ALLHAT	Antihypertensive and Lipid Lowering Treatment to Prevent Heart Attack Trial
AMA	American Medical Association
APPROVe	Adenomatous Polyps Prevention on Vioxx
BDA	Bayh-Dole Act
BIO	Biotechnology Organization
CAM	Complementary and alternative medicine
CDA	Confidential disclosure agreement
CLA	Cross-license agreement
COC	Conflict of commitment
COD	Conflict of duty
COI	Conflict of interest
COPR	Council of Public Representatives
CRADA	Cooperative research and development agreement
CRO	Contract research organization
CV	Curriculum vitae
DNA	Deoxyribonucleic acid
DOE	Department of Energy

ECU	East Carolina University
ELSI	Ethical, legal, and social issues
EPA	Environmental Protection Agency
ES	Embryonic stem (cells)
FDA	Food and Drug Administration
FFP	Fabrication, falsification, and plagiarism
GDP	Gross domestic product
HGP	Human Genome Project
HIV	Human immunodeficiency virus
IACUC	Institutional Animal Care and Use Committee
ICMJE	International Committee of Medical Journal Editors
IND	Investigational new drug
IPRs	Intellectual property rights
IRB	Institutional Review Board (for human subjects research)
ISRCTN	International Standard Randomized Controlled Trial Number
JAMA	*Journal of the American Medical Association*
LDL	Low-density-lipoprotein
LPU	Least publishable unit
MIT	Massachusetts Institute of Technology
MTA	Material transfer agreement
NAS	National Academy of Sciences
NASA	National Aeronautic and Space Administration
NCBI	National Center for Biotechnology Information
NEJM	*New England Journal of Medicine*
NIEHS	National Institute of Environmental Health Sciences
NIH	National Institutes of Health
NIMH	National Institute of Mental Health
NSAID	Non-steroidal anti-inflammatory drugs
NSF	National Science Foundation
ODA	Orphan Drug Act
OHRP	Office of Human Research Protection
ORI	Office of Research Integrity
PCMH	Pitt County Memorial Hospital
PDUFA	Prescription Drug User Fee Act
PhRMA	Pharmaceutical Research and Manufacturers of America
PHS	Public Health Service
PLoS	Public Library of Science
PRIM&R	Public Responsibility in Medicine and Research

R&D	Research and development
RCT	Randomized controlled trial
REB	Research Ethics Board, Toronto General Hospital
REC	Research Ethics Committees (for human subjects research)
RTLA	Reach-through license agreement
SSRI	Selective serotonin reuptake inhibitor
TGH	Toronto General Hospital
TIGR	The Institute for Genomic Research
TTA	Technology Transfer Act
UCLA	University of California, Los Angeles
USPTO	U.S. Patent and Trademark Office
UT	University of Toronto
VIGOR	Vioxx Gastrointestinal Outcomes Research
WARF	Wisconsin Alumni Research Foundation

THE PRICE OF TRUTH

ONE

SCIENCE AND MAMMON

> The love of money is the root of all evil: which while some have coveted after they have erred from the faith, and pierced themselves through with many sorrows.
>
> —1 Timothy 6:10 (King James Version)

1.1 Introduction

Modern science is big business. Like any other business, science is influenced by economic forces and financial interests (Greenberg 2001). Three sectors of our economy with very different interests—academia, government, and private industry—spend billions of dollars each year on research and development (R&D). Private corporations use scientists to develop new products and innovations. Governments use scientists to inform public policies, promote the general welfare, and develop military technologies and tactics. Academic institutions use scientists to advance knowledge and learning, enhance their reputations, and generate revenue. These patrons of science have different visions, vices, and virtues, but they are all willing to spend a great deal on science. In serving these different masters, do scientists risk losing their most cherished values? Can money, regardless of its source, corrupt science? This book is about how money can adversely affect the conduct of research, and how scientists and the public can take steps to safeguard the norms and standards that constitute the scientific mode of thought. Before embarking on this exploration, it will be useful to review some economic trends related to modern science and some cases studies on the relationship between science and money, so that the reader may have a better understanding how market forces and financial interests operate in science.

1.2 The Economics of Modern Science

The United States government's investment in R&D has grown steadily since World War II (Greenberg 2001). The U.S. government budgeted $132 billion for research and development (R&D) in 2005, which included $71.2 billion in defense R &D and $60.8 billion in civilian R&D. The NIH, at $28.8 billion, took the largest share of civilian R&D, followed by the National Aeronautics and Space Administration (NASA) at $16.2 billion, the National Science Foundation (NSF) at $5.7 billion, and Department of Energy (DOE) at $3.4 billion (Malakoff 2004).

Even though U.S. government spending on R&D has been enormous, private spending on R&D has outpaced government spending since the 1980s. In 2004 in the United Sates, private industry spent more than $200 billion on R&D, nearly double the federal government's contribution (see table 1.1). The pharmaceutical industry is one of the leading sponsors of R&D. In 2002, companies belonging to the Pharmaceutical Research and Manufacturers of America

Table 1.1 U.S. Funding of R&D by Source (not adjusted for inflation)

Year	Total R&D*	Federal**	Non-Federal***
1952	5.0 (est.)	2.7	2.3
1956	8.5	5.0	3.5
1960	13.7	8.9	4.8
1964	19.1	12.8	6.3
1968	24.7	15.0	9.7
1972	28.7	16.0	12.7
1976	39.4	20.3	19.1
1980	63.3	30.0	33.3
1984	102.3	46.5	55.8
1988	133.9	60.1	73.8
1992	165.4	60.9	104.5
1996	197.3	63.3	134.0
2000	264.6	83.3	181.3
2004	368.1	122.7	245.4

*In billions of current U.S. dollars not adjusted for inflation.

** Source of funding is the federal government. Amount includes both defense and civilian research.

***Sources of funding other than the federal government, such as private industry (the dominant source), private foundations, and universities and colleges.

Source: Based on National Science Foundation (2002), Greenberg (2001), Shamoo and Resnik (2003), and Malakoff (2004).

(PhRMA) spent $32 billion on R&D (Pharmaceutical Research and Manufacturers of America 2003). In 2001, biotechnology companies belonging to the Biotechnology Industry Organization (BIO) spent $15.7 billion on R&D (Biotechnology Industry Organization 2003). Total funding for biomedical R&D in the United States increased from $37.1 billion in 1994 to $94.3 billion in 2003. Industry-sponsored clinical trials increased from $4.0 billion in 1994 to $14.2 billion in 2004 (Moses et al. 2005). Although most people think of scientists as working in university laboratories, over 70% of R&D in the United States is conducted at private labs (Jaffe 1996). Private industry sponsors about 10% of research conducted in academic institutions (Bok 2003).

Scientific and technical research has a significant impact on the U.S. economy. The United States sponsors 40% of the world's R&D, and between six and eight million scientists in the United States are employed in research and development (Greenberg 2001). The U.S. gross domestic product (GDP) was estimated at $9.5 trillion in 2003 (Bureau of Economic Analysis 2003). If one projects past spending trends to 2003, the United States spent over $300 billion on R&D in 2003, which constitutes about 3.2% of the U.S. GDP. Because spending on R&D helps to create a variety of jobs not directly related to R&D, the total economic impact of scientific and technical research is probably much greater than 3.2% of the U.S. economy. For example, in the United States in 1999, the biotech industry created 437,000 jobs, which included 151,000 direct jobs and 287,000 indirect jobs (Biotechnology Industry Organization 2003). If spending on R&D results in the same percentage of jobs created in other industries, then economic activity related to R&D would account for about 6% of the U.S. GDP. R&D investments have a similar impact on other economies of other developed nations, such as Japan, Germany, France, and the United Kingdom (U.K.) (May 1999). (See table 1.2). Additionally, the new technologies that result from R&D investments generally enhance overall economic productivity and development: the countries with the largest investments in R&D also have the strongest economies.

The large sums of money at stake in scientific research have had an impact on individual scientists and universities. Although most scientists have aspirations that are more honorable than the desire for wealth, it is naïve to think that money has no influence on the behavior of scientists. While few scientists are millionaires, many earn a sizeable income, well above the U.S. median household income

Table 1.2 R&D Expenditures as a Percentage of Gross Domestic Product

Year	U.S.	Japan	Germany	France	U.K.	Italy	Canada	Russia
1990	2.62	2.85	2.75	2.37	2.16	1.29	1.47	2.03
1993	2.49	2.68	2.35	2.40	2.12	1.13	1.63	0.77
1996	2.53	2.80	2.26	2.30	1.91	1.01	1.60	0.90
1999	2.63	3.01	2.38	2.17	1.87	1.04	1.58	1.06

Based on National Science Foundation (2002).

of $63,228 (U.S. Census Bureau 2003). The median income for an experienced biostatistician with a PhD in the United States in 2003 was $158,609; followed by physicist, $132,711; geologist, $118,184; astronomer, $115,925; and chemist, $111,582. Clinical researchers with an MD or an MD/PhD have salaries in the $200,000 to $350,000 range, depending on their medical specialty. For comparison, the median income for an experienced high school teacher in the United States in 2003 was $63,571 (Salary.com 2003). Scientists who work in the natural and biomedical sciences and engineering tend to earn much more money than their colleagues in the humanities and social sciences (Bok 2003; Press and Washburn 2000). A study of fifteen metropolitan universities conducted by the University of Louisville showed that faculty in the natural sciences, basic biomedical sciences, and engineering earn, on average, $10,000 to $30,000 more a year more than their colleagues in the humanities and social sciences (University of Louisville 2002).

Many researchers, especially those in the biomedical sciences, supplement their income with financial arrangements related to their research, such as ownership of stock, consulting contacts, honoraria, and royalties from patents (Bodenheimer 2000). A national survey of 2,052 life sciences faculty found that 28% received research support from industry (Blumenthal et al. 1996). Another survey found that 34% of lead authors had financial interests related to their research; 20% served on advisory boards of companies; 7% were officers, directors, or major share holders; and 22% were listed as inventors in a related patent or patent application (Krimsky et al. 1996).

Some scientists, especially post-doctoral researchers, do not earn a decent income or have good job security. Post-doctoral researchers are scientists who have completed their dissertation but have not obtained an academic or an industry job. At universities, they are usually paid from research grants or contracts obtained by senior investigators. They are "soft money" employees because their economic livelihood depends on dollars from grants or contracts. At universities, post-doctoral students earn $25,000–$35,000 a year, with no fringe benefits. The ranks of post-doctoral researchers have grown steadily as employers have exploited cheap sources of scientific labor (Greenberg 2001).

Some surveys have found that financial interests in research are having an effect on data sharing and publication in academic science. In one survey, 19.8% of life scientists reported that they had delayed or stopped publication in order to protect pending patents or other proprietary interests, to negotiate license agreements, or to resolve intellectual property disputes (Blumenthal et al. 1997). In a survey of 1,849 geneticists, 47% of respondents said that they had been denied at least one request for data, information, or materials in the last three years, and 35% of those surveyed said that sharing of data had decreased in the last decade (Campbell et al. 2002).

Private companies usually require their researchers to delay or withhold publication in order to protect patents or trade secrets. A survey of biotechnology companies conducted in 1994 found that 82% of the firms require scientists to sign confidential disclosure agreements (CDAs). Of the firms with CDAs, 88% said that these agreements also apply to students. The survey found that 47% of the companies require confidentiality in order to protect patents. In addition, 41% of the companies in the survey said they the derive trade secrets from academic research that they sponsor (Blumenthal et al. 1986). According to a different survey, 53% of scientists at Carnegie Mellon University had signed contracts that allowed companies to delay publication (Gibbs 1996). With private companies, researchers also sign material transfer agreements (MTAs) that prohibit them from sharing research materials, such as reagents or cell-lines, without the company's permission.

A growing body of evidence supports a strong connection between the source of funding and the content of published results (Krimsky 2003; Davidson 1986; Angell 2004; Goozner 2004). The majority of published studies sponsored by private companies tend to favor the companies' products. One study found that 100% of articles reporting positive results for non-steroidal, anti-inflammatory drugs favored the companies that sponsored the research (Rochon et al. 1994). Another study found that 98% of drug studies published in a symposium proceedings favored the sponsoring company (Cho and Bero 1996). According to another study, only 5% of industry-sponsored articles on drugs used in cancer treatment reported negative findings, as opposed to 38% for non-industry-sponsored research (Friedberg et al. 1999).

One of the most impressive studies illustrating the connection between funding and results analyzed seventy articles on calcium channel blockers (drugs used to treat hypertension) published in medical journals from 1995 to 1996. The study found that 96% of the authors who published research supporting the use of calcium channel blockers had financial relationships with pharmaceutical companies that manufacture these drugs. Only 37% of the authors who published research that questioned the use of calcium channel blockers had such relationships (Stelfox et al. 1998).

There are several reasons why researchers, especially biomedical scientists, have financial interests in research. First, many scientists are employed by private companies or have contracts or grants to conduct research for private companies. Second, many scientists now own intellectual property related to their research, such as patents. Third, scientists have formed their own start-up companies to transform the fruits of their research into marketable products. Because start-up companies often have a limited amount of capital, they often make payments in the form of shares of stock. For example, a researcher could invest his/her time in the new company in return for shares of stock or stock options.

Many universities are now engaging in commercial activities related to research. During most of the twentieth century, universities did not take a great interest in patenting. Indeed, many university faculty and administrators regarded patenting as either immoral or against the interests of science, and some put pressure on faculty members not to seek patents (Bok 2003). In the 1920s, the Board of Regents from the University of Wisconsin refused to accept any funds from Henry Steenbock's patent on a process for irradiating milk. Some of Wisconsin's alumni formed a nonprofit research foundation, the Wisconsin Alumni Research Foundation (WARF), to fund research. After accepting Steenbock's patent, WARF went on to generate millions of dollars through thousands of patent disclosures and hundreds of license agreements (Bowie 1994). Herbert Boyer and Stanley Cohen, who discovered how to use bacteria to splice genes, initially did not seek a patent on this new technology. They eventually donated their patent to the University of California at San Francisco, which collected more than $150 million in royalties from Boyer and Cohen's invention (Bok 2003).

Universities and colleges abandoned their ambivalence toward the commercialization of research in the 1980s as a result of several factors, including the need for alternative sources of revenue because of declining state funds, the adoption of a business mentality in academia, the creation of closer ties with industry leaders, and the passage of the Bayh-Dole Act (BDA) in 1980, which allowed scientists to patent research that was developed with government funds (McPherson and Schapior 2003; Shamoo and Resnik 2003). The purpose of the BDA was to encourage universities to help develop useful technologies and transfer them to the private sector. Congress passed the Technology Transfer Act (TTA) in 1986 to allow the federal government to negotiate cooperative research and development agreements (CRADAs) and material transfer agreements (MTAs) with private companies and universities to transfer technology from the public to the private sector (Shamoo and Resnik 2003).

The BDA and the TTA appear to have succeeded in achieving their goals. From 1991 to 2000, the amount of money universities earned from royalties increased by 520% and the number of patent applications increased by 238% (Thursby and Thursby 2003). In 2001, colleges and universities filed 9,454 U.S. patents, signed 3,300 licenses, and earned $827 million in royalties and other payments from patents (Blumenstyk 2003), which was slightly down from the all-time high of $1 billion set in 2000. In 2002, the University of California was the leading academic patent holder (431 patents), followed by the Massachusetts Institute of Technology (135 patents), and the California Institute of Technology (109 patents) (*Science* 2003). Although $1 billion is a lot of money from patents, the amount is still far less than other sources of university revenue. For example, in 2001, colleges and universities received $24.2 billion in private donations from individuals and corporations (Strom 2002).

In 1980, only a handful of universities had technology transfer offices. Today, all major research universities and government laboratories have technology transfer offices (Bok 2003). Technology transfer offices help researchers convert their scientific discoveries into patentable inventions; help researchers identify useful products and take charge of the patent application process; and promote the licensing of patented technologies to private corporations for commercial development. Universities usually also require researchers to disclose patentable inventions and to assign control of these inventions to the university. Universities, like private companies, claim to own all patentable inventions and commercial products developed in their laboratories or with their funds. In return, universities usually provide scientists with a share of royalties from patents, typically 50% or less (Shamoo and Resnik 2003). Universities usually take their share of royalties from patents and reinvest that money in technology transfer operations or research activities.

Universities have also developed close ties with industry through partnerships with corporations and restricted, corporate gifts (Angell 2004; Krimsky 2003). The first major corporate investment in a university occurred in 1974, when Harvard University and Monsanto signed a $23 million agreement in which Monsanto provided Harvard with money for research in exchange for the right to patent all of the products of research. Another influential deal took place in 1981, when the Massachusetts Institute of Technology (MIT) and Technicon forged a $120 million deal to form The Whitehead Institute (Shamoo and Resnik 2003). Other universities have followed these examples. For instance, Lilly Foundation, which was established by the drug-maker Eli Lilly, recently made a $105 million gift to Indiana University for the purpose of promoting research and education (Strom 2002). In 2002, the Walton family, who own Wal-Mart, pledged $300 million to the University of Arkansas for improving the honors program and graduate school (Pulley 2002).

One of the most unusual university-industry deals occurred in 1998, when the regents of the University of California at Berkeley signed a five-year, $25 million agreement with Novartis Agricultural Discovery Institute to fund research in the Department of Plant and Microbial Biology. All of the faculty members from the department were allowed to sign the agreement, which provided the department with $3.3 million a year in unrestricted research money and $1.67 million a year in administrative costs. In exchange for its investment, Novartis was granted the first right to negotiate a license from discoveries made using its funds. A committee composed of members from the Department of Plant and Microbial Biology and Novartis decided how the R&D funds would be allocated (Krimsky 2003).

Universities are also increasing the amount of money they receive from private industry to conduct R&D. During the 1990s, the University of Texas at Austin increased the amount R&D funding it receives from industry by 725%. It was followed by University of California at San Francisco, 491%; Duke University, 280%; and Ohio State University, 272%. Yet overall R&D funding at these institutions grew only from 13% to 85% during the 1990s. By 2000, 31% of Duke's R&D budget came from industry. It was followed by the

Georgia Institute of Technology, 21%; the Massachusetts Institute of Technology, 20%; and Ohio State University, 16% (Krimsky 2003).

It is worth noting that universities are losing some of their clinical trial business to physicians in private practices. Currently, over 60% of clinical trials are community-based studies conducted through physician's offices or clinics, while less than 40% of clinical trials are university-based studies conducted through academic health centers. Private companies known as contract research organizations (CROs) help to organize and implement community-based studies. The CROs provide research sponsors with access to physicians, patients, and hospitals, and they help with patient recruitment, data auditing, and research oversight (Angell 2004). The percentage of community-based studies has tripled in the last decade (Morin et al. 2002). Because clinical trials are an important source of funding, academic health centers are taking steps to regain their share of the market, such as improving their relationships with pharmaceutical and biotechnology companies and making human research review and oversight more reliable and efficient.

Universities have also taken an active role in helping faculty to form start-up companies. One of the biggest difficulties in forming a new company is obtaining capital from banks (in the form of loans) or investors (in the form of stock purchases). Many universities have helped their faculty-entrepreneurs to overcome these financial difficulties by investing funds in the new company in exchange for stock or equity in the company. In a typical arrangement, a faculty member develops a new technology, patents it, assigns the patent to the university, and then forms a private company to develop the technology. The university licenses the technology to the private company and also provides it with start-up funds. Many universities have formed private, research foundations to hold investments in start-up companies, make investments in other companies, and to distribute venture capital to researchers who are forming new companies (Krimsky 2003).

Money from government contracts and grants also affects scientists who do not obtain funding from private industry. Scientists who do not have support from industry must obtain funding from government contracts or grants to continue their research and advance their careers. Even though the United States continues to increase government support for R&D, competition for grants and contracts is highly competitive because funding is always scarce. Funding is perpetually scarce because the demand for funding continues to exceed the supply, due to the glut of scientists in the labor force and unlimited desire for more knowledge (Greenberg 2001). Scientists who are not able to acquire funding from public or private sources will soon have to find a different type of work. In science, funding is a meal ticket. The need to obtain funding can place financial pressures on scientists, and entices some to fabricate or falsify data or stretch the truth on grant applications and reports to granting agencies. Funding pressures can also affect the fairness of the peer review system and encourage scientists to exploit the system in order to obtain funding (Shamoo and Resnik 2003).

Universities and colleges also have responded to the lure of government money. Revenue from government contracts and grants constitutes a significant source of income for a typical research university (Bok 2003). (See table 1.3.) In 1998, colleges and universities received $25 billion in government grants or contracts (Greenberg 2001). Because government funding of R&D rose by 50% 1998 to 2003, colleges and universities probably received about $37 billion from government contracts or grants in 2003.

Table 1.3 Top Twenty Academic Institutions, Ranked by R&D Expenditures, 1999*

University of Michigan, all campuses	$509 million
University of Washington–Seattle	$483 million
University of California–Los Angeles	$478 million
University of Wisconsin–Madison	$463 million
University of California–San Diego	$462 million
University of California–Berkeley	$452 million
Johns Hopkins University	$439 million
Johns Hopkins University, Applied Physics Lab	$436 million
Stanford University	$427 million
Massachusetts Institute of Technology	$420 million
University of California–San Francisco	$417 million
Texas A&M University, all campuses	$402 million
Cornell University, all campuses	$396 million
University of Pennsylvania	$384 million
Pennsylvania State University, all campuses	$379 million
University of Minnesota, all campuses	$371 million
University of Illinois–Urbana–Champaign	$358 million
Duke University	$348 million
Harvard University	$326 million
Ohio State University, all campuses	$323 million

*Sources of funding include federal government, state/local government, private industry, and other sources. Based on NSF (2002).

Government contracts or grants include direct costs, which pay for salary, students, equipment, tools, and so on, as well as indirect costs, which cover administrative costs associated with research. Government agencies negotiate indirect costs with universities. While there is some variation in the amount of indirect cost rate, the typical rate is 30%. A grant for $100,000 in direct costs, for example, would also provide the institution with $30,000 to cover indirect costs. Income from indirect costs is a very useful source of revenues for university and college administrators, since they can use this money to pay for virtually anything that might support the institution's research infrastructure. Although most indirect costs support expenditures necessary for research, such as libraries or administrative support for research, much money is leftover for other uses. Indirect costs are academia's slush fund (Greenberg 2001).

Another increasingly common source of government funding comes from earmarked money. Earmarked money bypasses the normal route of disbursement, which employs peer review panels in decision-making. Congressional appropriations are the usual source of earmarked funds. For instance, a U.S. congressman might push to have funding for a research project for his home district attached to an omnibus appropriations bill. The government pays for many of these pork barrel projects every year. Most congressmen and congresswomen subscribe to an unwritten rule for pork barrel projects: you endorse my pork and I'll endorse your pork. Negotiations for pork barrel projects often take place behind closed doors. In 1999, universities and colleges lassoed $797 million in earmarked funds. The top twenty research universities usually pull in more than half of this amount. When one includes government laboratories, the total amount of earmarked money is about $6 billion annually (Greenberg 2001).

Because government funding is such an important source of revenue, most universities and colleges make personnel decisions about scientists based on their ability to draw R&D dollars to the institution (Bok 2003). Scientists who get contracts and grants can receive tenure, a promotion, a salary increase, and other rewards. Scientists who do not get contracts and grants lose out. It has been said in academia that publication is a meal ticket (LaFollette 1992). In science at least, it is just as important (if not more important) to pull in money from contracts or grants.

Indeed, universities now frequently compete to hire nationally known scientists in order to acquire not only their talent and reputation but also their projected grants and contracts. They often entice scientists with offers of a higher salary, more autonomy, or better facilities or equipment. For example, when Randolph Chitwood, a cardiac surgeon at the Brody School of Medicine at East Carolina University (ECU) who is one of the world's leading experts in robotic surgery, received an offer from Harvard Medical School, ECU and Pitt County Memorial Hospital (PCMH) responded by increasing his income from $672,834 a year to $950,000 a year and promising to build a $60-million cardiac center and name him the director. Chitwood brings in about $1.85 million a year to ECU in grants and contracts and draws many patients and surgical trainees to PCMH. Duke University wooed two pioneering photonics researchers, David and Rachel Brady, away from University of Illinois at Urbana-Champaign when Duke's engineering school announced that it had received a $25-million gift to build a center for photonics and communications systems. Chemistry professor Joseph Desimone turned down offers from the Georgia Institute of Technology and the University of Florida when the University of North Carolina at Chapel Hill and North Carolina State University agreed to establish a new Triangle National Lithographic Center and buy a $3-million lithography instrument (Bonner 2003).

Many commentators are concerned that this emphasis on funded research can have a negative impact on other academic missions, such as teaching (and mentoring), academic advising, and public service. Few universities make a serious effort to evaluate teaching, improve teaching, or reward professors for good teaching. For most universities and many colleges, the ability to perform funded research is the most important factor in assessing faculty (Bok 2003).

1.3 Science and Money: Case Studies

The previous section presented some facts concerning the economics of science. To show how these market forces and financial interests can impact scientific research, it will be useful to ponder some cases from the recent history of science. Although one might argue that case studies provide anecdotal rather than statistical evidence, one cannot deny the rhetorical and educational value of salient examples. Subsequent chapters will also refer to these examples in discussing the impact of money on science.

The Race to Sequence the Humane Genome: Public versus Private Science

The race to sequence the human genome epitomized the opportunities and the dangers of investing huge sums of public and private money into scientific research. In October 1990, the U.S. government launched the Human Genome Project (HGP), a $3-billion effort to map and sequence the human genome in fifteen years. The DOE, looking for scientific and political relevancy in the post–Cold War era, collaborated with the NIH to hatch and nurture the HGP. The two rival agencies established the International Human Genome Sequencing Consortium to accomplish and carry out the goals of the project. James D. Watson, who shared the Nobel Prize with Francis Crick in 1953 for discovering the structure of deoxyribonucleic acid (DNA), was the initial director the of HDP. Watson left in 1992 and was replaced by Francis Collins (Human Genome Project 2003a).

In May 1998, J. Craig Venter and Applera Corporation founded Celera Genomics, a private company whose primary mission was to sequence the entire human genome in three years (Celera 2003). Venter, who had become frustrated with the bureaucracy at the NIH, left that agency in 1991. In 1992, he founded a nonprofit company, the Institute for Genomic Research (TIGR). Venter collaborated with Michael Hunkapillar, head of scientific equipment at PE Biosystems, to assemble a large cache of automated DNA sequencing machines. Venter planned to use his novel "shotgun" approach, which the NIH had rejected, to sequence the human genome. Instead of tediously sequencing the genome one piece at a time, Venter had the idea to break apart the entire genome, sequence the pieces, and put the pieces back together using algorithms run on supercomputers to match the pieces (Service 2000). Although scientists who followed the traditional "clone by clone" method were skeptical about the quality and reliability of the "shotgun" approach, Venter silenced his critics in February 2000 when Celera published a high quality sequence of the fruit fly (*drosophila melanogaster*) genome.

By June of 2001, Celera had sold over $1 billion in stock and had spent $300 million on sequencing the human genome. Celera planned to profit from its R&D investments by selling early access to human genomic data to pharmaceutical and biotechnology companies. It also planned to patent a limited number of genes. Celera had to revise its business plan when it was unable to generate among private firms substantial interest in early access to genomic data. Celera has gone back to a more traditional business strategy: patenting macromolecules and providing researchers and corporations with access to genomic databases and information services (Resnik 2003a).

The period from 1998 to 2001 marked a time of cooperation and competition between the public and private efforts to sequence the human genome. On one hand, researchers from the public and private sectors worked together to develop DNA sequencing technologies, tools, and techniques. They also shared data, ideas, and results. On the other hand, publicly funded and privately funded researchers had philosophical differences concerning data access and release policies. The International Human Genome Sequencing Consortium was committed to free and rapid release of DNA sequence data prior to publication (Collins, Morgan, and Patrinos 2003). Researchers from the consortium deposited their data in GenBank, a public database for genetic sequence data. Celera, however, did not accept free and rapid release of DNA sequence data, and its researchers did not deposit data in GenBank. Celera planned to charge corporations or research institutions a fee for an early look at genetic data, although it eventually granted researchers free access to data through its Web site. In return for access to data, researchers were required to agree to not commercialize the data or distribute it publicly (Marshall 2001a).

The two sides eventually reached an agreement (of sorts), and they published their own draft versions of the human genome on February 16, 2001, two years ahead of the HGP's original deadline for having a complete draft of the human genome. The competitive atmosphere created by the rivalry between Celera and the consortium, as well as the infusion of millions of dollars of private funds into this effort, helped to accelerate the pace of research. The consortium published its results in *Nature*; Celera published its results in *Science*. Some researchers objected to *Science*'s decision to publish Celera's results without requiring Celera to deposit its data in a free, publicly accessible Web site. By April 2003, Celera and the consortium had completed their efforts and published final versions of the human genome.

Since genetics play an essential role in cell structure, regulation, and function, as well as in many disease processes, the HGP will probably have a significant impact on the biomedical sciences as well as clinical medicine. Some noteworthy fields that have already benefited from the HGP include gene therapy, oncology, reproductive medicine, pathology, and human genetics. The HGP also has provided a wealth of information that will be useful in genetic testing, screening, and fingerprinting. Although scientists now have a wealth of human genetic data, they still face the daunting task of determining what it all means (Collins and McKusick 2001). There are many different steps in the pathway from genotypes to phenotypes that scientists know very little about, such as how genes are regulated, how genes code for different proteins, and how proteins function in the body. The practical and scientific benefits of the HGP may still be decades away.

The HGP was controversial from its inception. Many scientists were concerned that the project would divert government funds from other worthy research programs. Since public funds are limited, government science funding is a zero-sum game: money given to one discipline or project takes money away from another. Some scientists, especially those from the physical and social sciences, questioned the fairness of investing a sum of money in one biomedical research project. Other scientists were concerned that the HGP would generate reams of useless data, since the goal of the project was to sequence all of the DNA in the human genome, including "junk DNA"—DNA sequences that do not code for proteins but which make up 98% of the genome. Finally, some scientists questioned whether the project could be done at all (Roberts 2001). While this objection seems mistaken from the present vantage point, it was plausible in 1990 when DNA sequencing labs could sequence only 500 sequences a day. Since the human genome has three billion basepairs, at that rate it would have taken eighty-two years to sequence the human genome with twenty labs.

However, the biggest controversies related to the HGP have been and continue be ethical, political, social, and legal. The knowledge gained from the HGP has significant implications for genetic testing, discrimination, and privacy; genetic patenting and commercialization; human reproduction, cloning, and eugenics; moral and legal responsibility; and our conception of human uniqueness and identity. These moral issues and problems existed before the HGP, of course, but the HGP has amplified and transformed them. For example, as a result of the HGP, it is likely that one day in the not-too-distant future it will possible perform a test on a person that will ascertain his or her complete genetic makeup, including genetic diseases, predispositions, and risk factors. Indeed, since leaving Celera in 2002, Venter has created another company that will sequence human, animal, or plant DNA for a fee. He hopes that in ten years he will be able sequence a person's entire genome for a few thousand dollars.

Because they realized the HGP would create significant ethical, political, and social controversies, the founders of the HGP had the foresight to devote 3%–5% of the project's annual funds to an ethical, legal, and social issues (ELSI) program. The ELSI program has funded dozens of grants, conferences, and workshops and is the world's largest bioethics program (Human Genome Project 2003b). The ELSI program has also helped to foster legislation related genetic discrimination and privacy and has developed a great deal of educational material on ethical, legal, and social issues for scientists, scholars, professionals, and the lay public.

The HGP was not the first "big" science project and it will not be the last. Some other notable civilian projects include the Hubble Telescope ($1.5 billion), Earth Observing System ($17 billion), and Space State Freedom ($30 billion) (Collins, Morgan, and Patrinos 2003). Defense-related science projects are bigger and even more expensive. For example, the United States has spent close to $100 billion on national missile defense research since President Reagan proposed the Strategic Defense Initiative in the 1980s (Marshall 2001b). The United States spent $8 billion on missile defense research in 2003 (Congressional Budget Office 2003). The Manhattan Project had a $2 billion price tag in 1945 dollars (or about $23 billion in 2003 dollars).

Suppression of Research

Three episodes from the 1990s illustrate some of the problems that can occur when academic researchers conduct research for private corporations. In 1995, the Boots Company forced Dr. Betty Dong, a clinical pharmacologist at the University of California at San Francisco, to withdraw a paper on medications used to treat hypothyroidism. The paper had been accepted by the *Journal of the American Medical Association*. Dong's research, which was funded by Boots, compared the company's medication, Synthroid, to several generic drugs. Dong found that the generic drugs were just as safe and effective as Synthroid, and that the United States could save millions of dollars a year if hypothyroid patients started using these generic drugs instead of Synthroid. Dong had signed a contract with Boots giving the company permission to review her results prior to publication. The contract also required Dong not to publish her work without the company's permission. Boots threatened to sue Dong and spent two years attempting to discredit her research. Dong withdrew the paper to avoid a lawsuit. Fortunately, the company relented, and Dong eventually published her results in the *New England Journal of Medicine* (Shamoo and Resnik 2003).

In 1994, David Kern, an occupational health physician working at Brown University Medical School and Memorial Hospital, examined a patient with shortness of breath who worked at the Microfibres plant. Kern decided to visit the plant to look for a cause of the patient's illness, but he could find none. In order to visit the plant, he signed a confidentiality agreement with the company. In 1995, Kern examined a second patient with the same problem who worked at the same plant. After reporting these incidents to the National Institute of Occupational Safety and Health, the company hired Kern to determine whether the work environment at the plant was causing this lung illness. Kern prepared an abstract on the illness for presentation at the 1997 meeting of the American Thoracic Society. Microfibres objected to his plan to present his work on the grounds that his research was premature and that he had signed a confidentiality agreement. Kern ignored the advice of a dean at Brown University and decided to publish the abstract because he believed that the confidentiality agreement he signed did not apply to his publication. A week after he presented his research at the meeting of the American Thoracic Society, he learned that his five-year joint appointment at Brown and Memorial Hospital would not be renewed (Krimsky 2003).

From 1993 to 1995, Dr. Nancy Olivieri and her collaborators at the University of Toronto (UT) and Toronto General Hospital (TGH) studied deferiprone, a drug used to treat thalassemia. They published an article on their research in the *New England Journal of Medicine* in 1995. Apotex, Inc., a Canadian pharmaceutical company that manufactures deferiprone, sponsored their research. Olivieri and her collaborators reported that the deferiprone was effective at reducing total body iron stores in patients with thalassemia and that the drug had manageable side effects. Shortly after she reported these positive findings, Olivieri observed that liver iron stores in many of her patients were reaching dangerous levels, which could cause heart failure or death. Olivieri decided to notify TGH's Research Ethics Board (REB) about this problem, so that patients could be warned about this new risk during the informed consent process. Apotex did not want Olivieri to report her concerns to the REB. When she notified the REB, the company terminated the study and withdrew all the supplies of deferiprone from the hospital's pharmacy. Additionally, the company threatened to bring a lawsuit against Olivieri if she decided to tell anyone else about her concerns about the drug. Olivieri continued to receive letters from the company threatening litigation, and she withdrew a presentation on deferiprone that she had planned to make at a scientific meeting (Olivieri 2003).

In 1998, Apotex was negotiating a large donation to UT and TGT. The company pressured both of these institutions to take action against Olivieri, and they responded to its demands. Both UT and TGT denied there was a problem with deferiprone, tried to repress awareness of the problem, sought to divide Olivieri from her colleagues, tried to discredit her work, and even attempted to have her dismissed from her position. In 1999, a group of scientists and ethicists intervened in this awful predicament and prevented Olivieri from being dismissed. Olivieri also reached an agreement with UT and TGT clearing her of all allegations of wrongdoing. A commission from the Canadian Association of University Teachers investigated the case in 2001 (Olivieri 2003).

Unpublished Results: Anti-Depressants and Vioxx

As mentioned earlier in this chapter, there is usually a strong connection between the source of funding and the research outcome of published research. One way that private companies can skew the published research record in their favor is to simply not publish unfavorable results. In the United States, pharmaceutical companies must obtain approval from the Food and Drug Administration (FDA) before they can market a new drug. To obtain FDA approval, a company must submit an investigational new drug (IND) application to the FDA and provide the organization with data about the safety and efficacy of the drug. The studies that generate this data must conform to a variety of regulations relating to the protections of human subjects and the objectivity and integrity of research design, data collection, and data analysis (Angell 2004; Goozner 2004). Pharmaceutical companies treat their research studies as trade secrets, and the FDA does not reveal these secrets to the public. Although the FDA publishes its findings concerning INDs, it does not publish data submitted by drug companies. Companies can decide to publish—or not to publish—data. Companies have a financial interest in publishing only favorable results from clinical trials, which some may regard as another form of marketing (Angell 2004). Only 30%–50% of studies sponsored by drug companies are published (Dickersin and Rennie 2003).

Scientists, physicians, and ethicists have known for many years about problems with unpublished data from clinical trials and have argued that all results from clinical trials should be published in journals or made available in a public database (Rennie 2000). Since the 1990s, various organizations have operated clinical trials registries, which provide information about clinical trials including enrollment criteria, objectives, and results of published and unpublished studies (Dickersin and Rennie 2003). For example, the NIH operates one of the largest registries, Clinicaltrials.gov. (www.clinicaltrials.gov). To date, clinical trial registration has been voluntary, not mandatory. As a result, many drug companies have decided not to register clinical trials.

In the spring of 2004, the biomedical community and the public learned about the dangers of prescribing anti-depressant drugs to children and adolescents. A systematic review published in *The Lancet* showed that all of the drugs classified as selective serotonin reuptake inhibitors (SSRIs), with the exception of fluoxetine (Prozac), were associated with an increased risk of suicide in children. Earlier, published studies had shown that SSRIs are effective at treating depression in children. However, *The Lancet* article included data from unpublished studies obtained by the Committee on Safety in Medicines, an advisory body to the British government, which showed that SSRIs are associated with an increased risk of suicide. The authors concluded that none of the SSRIs, except Prozac, should be used to treat depression in children or adolescents (Whittington et al. 2004).

The revelation that SSRIs can increase the risk of suicide in children and adolescents infuriated doctors, parents, researchers, and politicians. Government committees on both sides of the Atlantic investigated the problem of prescribing anti-depressants to treat depression in children and teens revealed through unpublished negative studies. New York Attorney General Eliot Spitzer sued GlaxoSmithKline, the manufacturer of Paxil, an SSRI, for fraud. The lawsuit claimed that the drug company had intentionally deceived doctors by failing to tell them some studies had show that the drug was not safe or effective at treating depression in children or teens (Harris 2004). The International Committee of Medical Journal Editors (ICMJE) decided to adopt a clinical trials registration policy. To be considered for publication in an ICMJE journal, a study must be registered in a public, clinical trials registry (DeAngelis et al. 2004). The U.S. Congress considered legislation to require all drug companies to register clinical trials. (As of the writing of this book, this proposed legislation had not yet been adopted.) Even the PhRMA, which represents the pharmaceutical industry in the United States, decided to create a voluntary registry (Couzin 2004).

Vioxx is the trade name of rofecoxib, a drug manufactured by Merck, which belongs to the class known as COX-2 inhibitors. COX-2 inhibitors impede the production of the COX enzyme that causes inflammation. COX-2 inhibitors provide pain relief and have a much lower risk of stomach bleeding or ulcers than other non-steroidal anti-inflammatory drugs (NSAIDS), such as aspirin or Motrin. In 1998, the FDA approved Celebrex, a COX-2 inhibitor manufactured by Pfizer, and in 1999 the FDA approved Vioxx. In 2000, Merck reported the results of the VIGOR (Vioxx Gastrointestinal Outcomes Research) study to the FDA. The VIGOR study showed Vioxx users had five times more heart attacks than users of Aleve, another NSAID but not a COX-2 inhibitor (Associated Press 2005). The VIGOR study, which was published in November 2000, reported the gastrointestinal benefits of Vioxx but did not mention the cardiovascular risks of the drug. Eleven out of twelve investigators in the VIGOR study had financial relationships with Merck, which sponsored the study (Bombardier et al. 2000). In 2001, the FDA warned Merck that it had misrepresented the safety profile of Vioxx during its promotional campaigns. In 2002, the FDA changed the warning label on Vioxx to reflect the cardiovascular risks found in the VIGOR study. In September 2004, Merck withdrew its drug, Vioxx, from the market after the APPROVe (Adenomatous Polyp Prevention on Vioxx) study showed that Vioxx could double the risk of a stroke or heart attack if taken for more than eighteen months. The APPROVe study was published in March 2005 (Bresalier et al. 2005). Merck started to defend itself from potentially thousands of lawsuits from Vioxx users, who had had heart attack or stroke while on the drug. In August 2005, the first trial concluded when the jury awarded $253 million to the widow of Robert Ernst, who used Vioxx. The jury found that Merck had been negligent because it failed to warn patients about the cardiovascular risks of the Vioxx. In November 2005, a jury found that Merck was not liable in Vioxx user Mike Humeston's heart attack (Associated Press 2005).

Tobacco Research and Litigation

In 1994, two scientists who worked at Philip Morris, Victor DeNobel, and Paul Mele, testified before Congress about secret research on the addictive properties of nicotine. The two scientists had both signed agreements with the company barring them from talking about their research or publishing any results without prior approval. The congressional committee that solicited their testimony released DeNobel and Mele from this agreement. DeNobel and Mele told the committee that they had been working in a secure, private laboratory on the addictive properties of nicotine and that tobacco companies were conducting research on nicotine in order to manipulate nicotine levels in cigarettes (Resnik 1998a). For years, tobacco companies had fought against additional FDA or congressional regulation by arguing that they were selling a natural product, not a drug. They had also denied that they were manipulating nicotine levels or attempting to make cigarettes more addictive. The FDA currently does not regulate tobacco as a drug, but many states and individuals have brought litigation against tobacco companies, including a $206 billion settlement between tobacco companies and forty-six states.

Conflict of Interest in Research

Eighteen-year-old Jesse Gelsinger's tragic death in a gene therapy experiment in 1999 is a classic example of how financial interests can cause (or appear to cause) bias in scientific research. Gelsinger suffered from a genetic liver disease, ornithine transcarbamase deficiency, which occurs when a person lacks a functional copy of gene that codes for the liver enzyme ornithine transcarbamase. Although people often die from this disease in infancy, Gelsinger was able to manage his disease with a controlled diet and drugs. Gelsinger agreed to participate in a Phase I human gene therapy study conducted by James Wilson and his colleagues at the University of Pennsylvania. Wilson had originally planned to conduct the experiment on very sick infants, but he decided to include Gelsinger in the study when other gene therapy researchers and bioethicist Art Caplan had convinced him that it would be unethical to perform the experiment on infants, who cannot give informed consent. The experiment used an adenovirus vector in an attempt to transfer functioning genes to Gelsinger's liver so that it could start producing the vital enzyme. Gelsinger died after developing a massive immune-response to the andenovirus, which had been infused into his liver (Marshall 2001c).

Federal investigators from the Office of Human Research Protection (OHRP) discovered some serious deviations from ethical and legal standards in this study. First, the researchers had not informed Gelsinger about toxic reactions to the adenovirus in other patients who had received the treatment, or about the deaths of several monkeys that had received the treatment. Second, the researchers had not reported these toxic reactions in human beings, otherwise known as adverse events (AEs), to the Food and Drug Administration (FDA) or the institutional review board (IRB) in charge of overseeing the research. Third, Wilson and the University of Pennsylvania both had significant financial interests in this research (Greenberg 2001).

Wilson, the principal investigator in the experiment, had twenty patents on several gene therapy methods and held 30% of Genovo's stock, a company that contributed $4 million a year to a nonprofit research organization established by the university, the Human Gene Therapy Institute, which sponsored the experiment. The University of Pennsylvania also held stock in Genovo. Wilson's 30% interest in Genovo violated Penn's policies that limited stock ownership by researchers conducting company-sponsored research to a 5% interest. Wilson had formed Genovo, as well as numerous other start-up companies, while at the University of Michigan. Penn lured him away from Michigan in 1993 by offering to name him as the director of the newly established Human Gene Therapy Institute. The fourteen-page informed consent document given to Gelsinger did include a section at the very end mentioning that Wilson and the University of Pennsylvania could benefit financially from successful experimental results (Greenberg 2001; Marshall 2001c).

Before Gelsinger's tragic death, several other notable conflict of interest (COI) cases caught the media's attention. In 1997, Michael Macknin conducted research on the effectiveness of zinc throat lozenges in treating the common cold. Shortly after he published this research, he bought stock in the company that manufacturers these lozenges, the Quigley Corporation, and he made a $145,000 profit when the price of the company's stock soared. Ironically, subsequent research published by Macknin showed that zinc lozenges are no more effective than a placebo in treating the common cold. Macknin sold his stock before publishing this negative result (Hilts 1997). In 1985, Harvard ophthalmologist Scheffer Tseng published a paper on an ointment containing vitamin A, which he used to treat dry eyes. Tseng did not disclose that he had obtained exclusive rights from the FDA to market the ointment for seven years. A drug company, Spectra, purchased these rights for $310,000. Tseng also did not disclose that he held 530,000 shares of Spectra. After he published his paper, the price of Spectra's stock soared, and Tseng sold his shares. Subsequent research showed the ointment was not as effective as these earlier studies indicated and that Tseng had minimized unfavorable results in his research. Tseng published his positive results before the negative results were revealed (Booth 1988).

Several episodes reported in the press have exposed conflicts of interest in the FDA's approval process (Krimsky 2003). Before approving any new product or method to diagnose, treat, or prevent a disease, the FDA requires the manufacturer of the new product or method to provide the agency with data from clinical trials designed to establish the safety and efficacy of the new product or method. The FDA requires that clinical trials abide by stringent rules for experimental design, data collection and monitoring, statistical analysis, adverse event reporting, and human subject protection. The FDA uses advisory committees composed of FDA scientists as well as outside consultants to review the data that the manufacturer submits to the FDA. In 1998, the FDA approved a vaccine, manufactured by Wyeth Lederle, used to prevent rotaviruses. One year after releasing the vaccine, Wyeth removed it from the market, because over one hundred children had developed bowel obstruction from the vaccine. A congressional committee investigating this incident found that the advisory committee that recommended approval of the vaccine included nine members with significant financial ties to companies that manufacture vaccines. One member of the committee owned $33,800 in stock from Merck, which develops vaccines. Another shared a patent on a rotavirus vaccine being developed by Merck and had received a $350,000 grant from Merck (Krimsky 2003).

An investigation of conflicts of interest in the FDA approval process, in the fall of 2000, examined financial interests in 159 meetings of eighteen FDA advisory committees. The investigation found that at least one committee member had a financial interest directly related to the topic being reviewed in 146 of the meetings. At 88 meetings, at least half of the committee members had financial interests directly related to the topic of the meeting. The financial interests of the committee members included grants, consulting fees, stock ownership, and patents (Krimsky 2003).

The FDA has its own conflicts of interests as well. Under legislation passed in 1992, the Prescription Drug User Fee Act (PDUFA), the FDA can charge fees to private companies for reviewing new products. The FDA can collect fees when a company submits an application for a new drug or biologic. In 2002, the FDA collected $143.3 million in fees from private companies. The FDA spent most of those fees on salaries and benefits for FDA employees. The FDA's total budget for salaries and benefits in 2002 was $1 billion (FDA 2002). Thus, the PDUFA fees represented 14% of the FDA's money that it spent on salaries and benefits. The purpose of the PDUSA is to improve the efficiency and quality of FDA review by enabling the FDA to hire additional staff. Prior to 1992, the FDA's funding came from taxes collected by the federal government. Some pharmaceutical industry watchdogs argue that the fees collected under the PDUFA make the FDA beholden to the drug industry (Wolfe 2003).

A recent study conducted by *Nature* has shown that many physicians who serve on expert panels that write clinical guidelines for prescribing drugs have connections to pharmaceutical companies. The study found that more than one-third of physicians on these expert panels had conflicts of interests. In one case, every member of the expert panel had received money from the company whose drug they were evaluating. Not surprisingly, the panel had a favorable opinion of the company's drug (Taylor and Giles 2005).

Another conflict of interest case worth mentioning happened over two decades ago. In 1976, John Moore received treatment for hairy-cell leukemia, a rare type of cancer at the University of California, Los Angeles (UCLA) Medical Center. David Golde, Moore's physician, recommended a splenectomy. After Moore's spleen was removed, Golde asked Moore to make additional visits to the Medical Center to provide blood and tissue samples. Without telling Moore, Golde used tissue from Moore's spleen as well as subsequent tissue samples to develop a cell line. Moore's tissue had scientific, medical, and commercial value because it was overproducing lymphokines, proteins that regulate a variety of immune-cell functions. Golde and his research assistant, Shirley Quan, signed agreements with several pharmaceutical companies and the University of California to develop the cell line. They also obtained patents on the cell line in 1984, which they transferred to the University of California (Resnik 2003b). The cell line eventually generated several billion dollars in income for a Swiss pharmaceutical company, which bought the patent (Krimsky 2003).

When Moore found out that Golde was commercializing his tissue without his permission, Moore sued Golde, Quan, the private companies, and the university for malpractice and conversion, namely, for substantially interfering with his personal property. The case made its way through the appeals process to Californian Supreme Court, which ruled that Moore could not prove the tort of conversion because he did not have a proprietary interest in the cell line derived from his tissues. The majority of the judges on the court held that

Moore did not have a proprietary interest in the cell line, but that the researchers who isolated, purified, and patented the cell line did have such as an interest. These judges defended this opinion on the grounds that allowing patients to have property interests in their own tissues would undermine biomedical R&D because companies would be unlikely to invest in developing products from tissues if they face legal uncertainties concerning their intellectual property claims arising from disputes with patients. However, the court also concluded that Golde committed malpractice because he failed to tell Moore about his commercial interests. The court ruled that Golde had a duty to obtain Moore's informed consent for the operation and the collection of tissue samples afterwards, which included an obligation to tell Moore about his commercial interests (*Moore v. Regents of the University of California* 1990).

In December 2003, a controversy over COIs in the intramural research program at the NIH erupted, following the *Los Angeles Times'* publication of several articles on consulting arrangements between NIH administrators and pharmaceutical and biotechnology companies (Willman 2003). The articles alleged that some NIH senior administrators had received hundreds of thousands of dollars from inappropriate consulting arrangements. NIH director Elias Zerhouni appointed a blue-ribbon panel in January 2004 to examine these charges and make policy recommendations. Two congressional subcommittees also held hearings on the topic and discovered other problematic relationships with drug companies, including dozens of researchers who had failed to report outside income. Harold Varmus, Zerhouni's predecessor, had loosened the NIH's ethics rules to recruit and retain top biomedical scientists encourage intramural researchers to collaborate with industry (Kaiser 2004a). In May 2004, the panel issued its report, which called for tighter controls on relationships with industry, and in July 2004, Zerhoni announced rules stricter than those recommended by the panel. Shortly after Zerhouni's announcement, the Office of Government Ethics issued a report recommending that the NIH prohibit all consulting with pharmaceutical companies (Kaiser 2004b). Zerhouni responded to this report by announcing a one-year moratorium on consulting with pharmaceutical or biotechnology companies for all NIH employees, to allow the NIH some time to change its policies and reporting procedures. In August 2005, Zerhouni announced the final ethics rules, which set limitations on the investments of NIH employees and prohibit some types of financial relationships with pharmaceutical companies, universities, trade associations, and other entities substantially affected by the NIH (NIH 2005).

The Cold Fusion Frenzy

In 1989, two scientists from the University of Utah, Stanley Pons and Martin Fleischmann, made an announcement that stunned the physics community: they declared at a press conference that they had discovered a method for producing nuclear fusion at room temperatures by running an electric current through an electrolyte solution. The two scientists made this announcement before they had published their results in a scholarly journal. Physicists from laboratories around the world scrambled to try to reproduce Pons and Fleischmann's astounding results. After they failed, most physicists concluded that there was no basis for cold fusion and that Pons and Fleischmann had probably made some simple mistakes in their interpretation of their experimental results. Pons and Fleischmann had succumbed to scientific self-deception. Other scientists were not as charitable and concluded that the two scientists had committed scientific misconduct. One reason that Pons and Fleischmann took the unusual step of reporting their results at a press conference, instead of in a scientific journal or at a scientific meeting, is that they wanted to establish that they were the first people to make this discovery; they wanted to protect their patent rights. By failing to disclose all the details of their experiment in the press conference, Pons and Fleischmann compounded problems with reproducing their results. The University of Utah also had invested a great deal of money in their research and hoped to share in the profit and the glory of cold fusion (Shamoo and Resnik 2003).

Following are two additional case studies concerning the relationship between science and money. While they do not deal with the corrupting influence of money, they do illustrate some problems that can arise when money is used to exert ideological or religious influence on science.

Politics and Peer Review

In October 2003, the House Energy and Commerce Committee held a hearing in which it asked the NIH to justify 198 approved or funded research projects dealing with HIV/AIDS prevention, risky behavior, pregnancy prevention, mental health, and other topics. The

Traditional Values Coalition, a conservative political group, gave the list of 198 grants to a staff member working for the House Energy and Commerce Committee. Because the NIH publicly announces all its research awards, the political group was able to learn about the grants through the NIH's public databases. The Traditional Values Coalition objected to many of the grants on the grounds that they endorse sexual behavior among teens and intravenous drug use. The purpose of these projects was to study methods of preventing HIV transmission through sex or intravenous drug use. Representative Henry Waxman (D-CA) called the list of 198 targeted grants "scientific McCarthyism." In 2003, the U.S. House of Representatives came within three votes of approving a motion, sponsored by Patrick Toomey (R-PA), to withdraw funding on four research grants on sexual behavior (Kaiser 2003). In the 1940s and 1950s, Alfred Kinsey faced similar hostility to his research on human sexuality. Kinsey funded his research through private funds, not government funds (Carey 2004).

In 2004, the House approved a bill to withdraw funding from two psychology grants; one of the grants was for a study of college students' perceptions of themselves, the other for a study of the relationship between dorm room décor and mental health. The bill also included a provision to limit travel to international meetings by NIH researchers (Kaiser 2004c). According to Representative Randy Neugebauer (R-TX), who sponsored the legislation, "Taxpayer dollars should be focused on serious mental health issues like bipolar disorders and Alzheimer's," not on "interior decoration" (Kaiser 2004c). Although the bill was not approved by the Senate, it established a precedent for challenging almost any grant in the social sciences and undermining the process of peer review by federal granting agencies.

Embryonic Stem Cell Research

Research on embryonic stem (ES) cells has been fraught with controversy ever since scientists discovered how to grow ES cell lines in vitro in 1999 (Thomson et al. 1998). There are many different types of stem cells in the body, each with different properties. Adult stem cells, in other words, stem cells found in the tissues of the body, are multipotent, which means that they can differentiate into multiple cell-types. For example, bone marrow stem cells can differentiate into red blood cells and several different types of white blood cells. Some types of ES cells are totipotent, which means that they can differentiate into any cell type or they can form a new organism, if implanted in the womb.

ES cells are extracted from the inner cell mass of early embryos known as blastocysts. Different types of stem cells may one day be useful in treating diseases that involve malfunctioning or destroyed cells or tissues, such as diabetes, heart failure, Alzheimer's dementia, and spinal chord injuries. Many researchers think that ES cells will be more useful than adult stem cells, since they can differentiate into many tissues types. Although it may be many years before researchers are ready to test ES cell therapies in humans, these treatments have yielded fruitful results in animals (Weissman 2002). The use of ES cells in research is controversial, since the process for deriving ES cells destroys the blastocyst, which many people consider to be a human being, with a right to life. Many abortion opponents are against ES cell research because it involves the destruction of the fetus (Green 2001).

In 2000, President Clinton authorized the use of NIH funds to study, but not derive, ES cells. The Clinton administration made this distinction to avoid violating the ban on the use of NIH funds for embryo research, which Congress enacted during the Reagan administration (Green 2001). This allowed NIH researchers to obtain ES cells from sources outside the NIH, such as biotechnology companies.

In 2001, President George W. Bush modified Clinton's policy and decided that the NIH would fund embryonic–stem cell research only on cell lines that had already been developed from embryos left over from attempted in vitro fertilization (Bruni 2001). The NIH-funded scientists could not conduct research on any new ES cell lines from any sources. At the time that Bush made this decision, it was believed that 64 cell lines were available, but only 21 cell lines are available to scientists from the NIH registry, as compared to about 130 cell lines worldwide (Daley 2004). Bush approved about $25 million dollars a year for ES cell research, as compared to about $300 million for adult–stem cell research. Bush also appointed the President's Council on Bioethics to study ethical issues related to ES cell research and other bioethics issues, such as cloning and genetic engineering (Holden and Vogel 2002). From the beginning, the council was embroiled in controversy. Critics of the council argued that Bush stacked it with members with conservative, anti-science views. The chair of the council, Leon Kass, has for many years been an outspoken opponent of reproductive technologies, including in vitro fertilization. In March 2004, Elizabeth Blackburn, a scientist, and William May, a theologian and ethicist, were asked to leave the council. Blackburn and May had both defended ES cell research. Bush replaced Blackburn and May with three members who oppose ES cell research (Blackburn 2004).

While Bush's decision to not invest U.S. government funds in ES cell research has hampered research in the United States, it has not stopped the research from occurring. Private companies in the United States, such as Geron Corporation and ES Cell International, have developed ES cell lines. Other nations, such as the United Kingdom, Sweden, Israel, Korea, and Singapore sponsor ES cell research (Holden and Vogel 2002). Universities in the United States, such as University of California at San Diego, the University of Wisconsin, and Harvard University, have established ES cell research institutes funded by private donors (Lawler 2004). Some states have also taken a stand on this research. For example, New Jersey and California have both passed legislation specifically permitting ES cell research. On November 2, 2004, California voters approved proposition to spend $3 billion on ES cell research over a ten-year period. Defenders of the amendment argued that it would promote important research not sponsored by the federal government and that it would help the state's economy by attracting biotechnology companies (D. Murphy 2004).

1.4 Overview of the Book

Scientists currently enjoy a great deal of influence, as governments, universities, and corporations have invested billions of dollars in scientific and technical research in the hope of securing power, prosperity, and profit. For the most part, this allocation of resources has benefited science and society, leading to new discoveries and inventions, disciplines and specialties, jobs, and career opportunities. However, there is also a dark side to the influx of money into science. The passage from the book of *Timothy* from the Holy Bible (quoted at the start of this chapter) conveys ageless wisdom about the consequences of greed for human affairs: the love of money causes people to stray from values and principles that they cherish. Although the author of Timothy is speaking to people of the Christian faith, his words apply equally to other traditions and practices, such as sports, education, art, medicine, politics, law, and science. Unbridled pursuit of financial gain can undermine not only religious piety but also fairness, trust, respect for human rights, democracy, and the search for truth.

The financial and economic aspects of modern science raise many important and difficult questions for research ethics, the philosophy of science, and science policy. How does money affect scientific research? Can financial interests undermine scientific norms, such as objectivity, truthfulness, and honesty? Have scientists become entrepreneurs bent on making money as opposed to investigators searching for the truth? How does the commercialization of research affect the public's perception of science? Can scientists prevent money from corrupting the research enterprise? What types of rules, polices, and guidelines should scientists adopt to prevent financial interests from adversely affecting research and the public's opinion of science?

This book will examine these and other questions related to the financial and economic aspects of modern science. It will investigate and analyze the relationship between the pursuit of financial gain and the pursuit of knowledge. It will consider how greed can affect the conduct of scientists, universities, and corporations, and the public's perception of science. It will explore how moral, social, and political values affect public and private funding of research, and how debates about research funding reflect broader, ideological divisions within society. The book will also propose some policies for controlling, regulating, and monitoring financial interests in research and for counteracting money's corrupting effects on science.

Some authors who have written on this these topics take a very dim view of the relationship between science and money (Press and Washburn 2000). According to Sheldon Krimsky:

> Public policies and legal decision have created new incentives for universities, their faculty, and publicly supported nonprofit research institutions to commercialize scientific and medical research and to develop partnerships with for-profit companies. The new academic-industry and non-profit–for-profit liaisons have led to changes in the ethical norms of scientific and medical researchers. The consequences are that secrecy has replaced openness; privatization of knowledge has replaced communitarian values; and commodification of discovery has replaced the idea that university-generated knowledge is a free good, a part of the social commons. Conflicts of interest among scientists has been linked to research bias as well as a loss of a socially valuable ethical norm—disinterestedness—among scientific researchers. As universities turn their laboratories into commercial enterprise zones and as they select their faculty to realize these goals, fewer opportunities will exist in academia for public-interest science—an inestimable loss to society. (*Science in the Private Interest*, 7)

Although I agree with Krimsky that money can have a corrupting influence on science, I also recognize that modern science is very different from the pure, "public-interest" science that he defends. If we take an honest look at the history of science, we will see that money has almost always had some affect on research, secrecy has always conflicted with openness, private knowledge has always coexisted with public knowledge, private-interest science has always occurred alongside public-interest science, and scientists have almost never been completely disinterested. Real science has been, and always will be, affected by various political, social, and economic biases, including private, financial interests (Hull 1988; Longino 1990; Resnik 1998a; Ziman 2000). The most prudent and realistic response to this situation is to try to mitigate the corrupting effects of private interests and to establish policies and procedures to safeguard the norms of science. Such rules should establish an appropriate balance between openness and secrecy, public ownership of research and private ownership of research, and public-interest science and private-interest science.

While this book will describe how money can affect science and will recommend some proposals for preventing money from corrupting science, it will not present an anti-industry diatribe on the evils of money, property, and commercialization, nor will it give a patronizing sermon on the intrinsic value of academic (or ivory tower) research uncorrupted by worldly pursuits. The book takes a realistic approach to this topic, not an idealistic one. The modern (i.e., developed, industrialized) world runs on principles of free-market economics. Even large countries that once opposed capitalism, such as Russia and China, have come to understand that free markets hold the key to economic growth. Although governments usually regulate markets to promote fair trade or protect social values, few institutions in the modern world can escape the influence of money. Science is no exception to this rule.

Modern science is big business because it has tremendous economic benefits and costs. It creates economic opportunities and entices economic risk-taking. Money turns the wheels of scientific and technical research and provides incentives for individuals and organizations. Scientists, universities, and corporations are agents participating in the free-market economy. They create, develop, modify, sell, buy, and license goods and services related to scientific and technical knowledge. Society cannot eliminate financial incentives and pressures from science any more than it can eliminate such pressures and incentives from sports, education, manufacturing, or health care.

Even a completely socialized science, funded and managed entirely by the government, would have its own economic (and social and political) biases and pressures. Financial interests play a role in science, whether one obtains money from a private corporation or the government. While the government usually does not have a vested interest in the outcome of research, the lure of government money affects scientific behavior much in the same way that the lure of private money affects behavior. Whether one seeks money from the government or a private company, one may be tempted to break ethical rules in order to gain financial rewards. Some researchers have lied on grant applications and on reports made to granting agencies in order to secure or maintain funding (Shamoo and Resnik 2003). Others have committed plagiarism or fabricated or falsified data on publications. The Office of Research Integrity (ORI), which deals with scientific integrity and misconduct involving research funded by the Public Health Service (PHS), reviewed 987 allegations of misconduct from 1993 to 1998 and closed 218 cases. Of the closed cases, 76 resulted in a formal finding of misconduct (Office of Research Integrity 1998).

Even though it is not realistic to think that one can liberate science from financial interests, one can still develop standards and regulations for managing and monitoring financial incentives and pressures that affect science. Governments, research organizations, universities, and corporations should adopt rules, policies, and guidelines for regulating the knowledge economy in order to promote economic fairness and to protect moral, social, political, and scientific values. This book will propose and discuss some rules and guidelines for safeguarding scientific research from the corrupting effects of money.

The plan for this book is as follows. The first part of this book will provide the reader with some historical, social, and philosophical background material that will be useful in understanding how money affects science. It will also articulate some concepts and theories that will be useful in understanding how financial interests and influences can affect science. Chapter 2 will explore the norms of science, including notions of objectivity, openness, truthfulness, respect for research subjects, and so forth. Chapter 3 will undertake a deeper explanation and justification of science's most important norm, objectivity. Chapter 4 will develop a framework for explaining how financial interests and influences can undermine the norms of science. The second part of the book, chapters 5 through 8, will apply the ideas developed in the first part of the book to particular topics and issues, such as intellectual property, conflicts of interest, scientific misconduct, scientific authorship and publication, and government support for research. The concluding chapter of this book, chapter 9, will make some policy recommendations based on the arguments and insights presented in chapters 1 through 8.

TWO

THE NORMS OF SCIENCE

Back off. I'm a scientist.

—Bill Murray, from the film *Ghostbusters* (1984)

2.1 Science and Values

Popular culture's image of science is schizophrenic. Some films and books depict scientists as detached, cold, brainy, laboratory researchers with little understanding of or interest in the humanities, arts, or human values. Others portray scientists as evil, calculating, reckless, and power-hungry geniuses with a perverted sense of themselves and human society. Still others represent scientists as courageous, wise, and humble heroes who save mankind from destructive monsters, deadly diseases, or out-of-control asteroids. All of these images are false and misleading.

Science is not value-free, nor is it inherently good or evil. Science, like any other human activity, has its own goals and principles, which constitute the norms or values of science (Merton 1973; National Academy of Sciences 1995; Haack 2003).[1] These norms apply to many different facets of scientific conduct, including experimental design, testing, confirmation, data analysis and interpretation, publication, peer review, collaboration, and education and training (Resnik 1998a). Scientists invoke norms to distinguish between well-designed and poorly designed experiments, good and bad theories, rigorous and shoddy methods, and appropriate and inappropriate behavior (Kuhn 1977; Longino 1990; Shrader-Frechette 1994).

There are at least two reasons why norms are essential to science. First, science is itself a type of a social activity composed of overlapping and interconnected communities (or societies). These different communities can be equated with different disciplines, such as biochemistry, quantum mechanics, immunology, economics, and so on. Scientific norms are necessary to promote cooperation and collaboration toward common disciplinary and interdisciplinary goals (Hull 1988; Longino 1990; Goldman 1999; Steneck 2004). Norms bolster trust among scientists and establish expectations. Science without norms is epistemological anarchy.

Second, science is practiced within a broader community (or society) and must therefore answer to that community's moral, social, and political norms: scientists can be and should be held publicly accountable (Shrader-Frechette 1994; Resnik 1998a; Goldman 1999). As members of society, scientists have a variety of moral obligations, namely, to tell the truth, keep promises, respect human rights and property, prevent harm, help others, obey the law, and participate in civic life. Like other professionals, scientists also have special obligations, based on their special role in society. The public entrusts scientists with authority, autonomy, and privileges. To maintain this trust, as well as the public's financial and political support, scientists should adhere to ethical standards (Steneck 2004).

2.2 The Goals of Science

To understand science's normative structure, it will be useful to distinguish between scientific principles (or standards) and scientific goals. Scientific principles are general rules that apply to scientific inquiry; scientific goals are the aims, objectives, or purposes of inquiry. There are two types of principles: categorical principles (or imperatives), which are justified without appealing to any goals, and teleological principles (or hypothetical imperatives) which are justified insofar as they promote particular goals (Kant 1753). Teleological comes from the Greek word "teleos" or "goal," and it means roughly "goal directed." One might argue that a rule such as "Wash your hands before eating" is teleological because it is justified provided that (a) one has the goal of avoiding infections spread by the hands and (b) washing one's hands is an effective way of achieving this goal. One might argue that a rule such as "Do not kill innocent human beings" is a categorical principle because it can be justified without referring to any goals. Science's principles are teleological rules, which are justified insofar as they promote the goals of science (Laudan 1984).[2] Scientific methods are rules (or sets of rules) for inquiry. Debates about scientific methods invariably appeal to the ability of those principles to advance scientific goals. For example, meta-analysis is a method for aggregating data from different clinical trials to evaluate the safety and efficacy of a particular medical test or treatment. Debates about the use of meta-analysis in medical research revolve around the ability of this method to yield reliable and truthful results. Those who support the use of meta-analysis argue that it yields reliable and truthful results; those who oppose its use argue that it does not (Bailar 1997). Adversaries in this debate agree that research methods should yield reliable and truthful results, but they disagree about whether meta-analysis can achieve this goal.

Sometimes scientists do not agree about goals or how to prioritize them when they engage in debates about scientific principles. Nevertheless, their debates still appeal to goals. For example, clinical researchers often use randomized controlled trials (RCTs) to determine whether a particular medical intervention is safe and effective. An RCT compares two or more treatment groups, such as an experimental treatment versus a standard treatment or an experimental treatment versus a placebo (a biochemically inactive substance). Some researchers believe that all new medications should be tested against a placebo in order to achieve reliable and statistically significant results, even when an effective therapy exists. Others object to using placebos when an effective therapy exists on the grounds that subjects in the placebo group will not receive an effective therapy that they would have received if they had not participated in the experiment. These researchers advocate the use of trials that compare experimental treatments to standard treatment, as opposed to trials that use placebos. The adversaries in these debates agree that clinical research should be reliable and statistically significant and should protect human subjects, but they disagree on the priority of these goals. Placebo advocates give reliability and statistical significance a higher priority than protecting human subjects, while placebo opponents reverse these priorities (Emanuel and F. Miller 2001). They both frame the debate as an argument about the goals of clinical research even though they disagree about how to prioritize those goals.

Since scientific norms are teleological, to understand scientific principles, one must understand the goals of science. It is important to distinguish between the goals of science and the goals held by individual scientists (Kitcher 1993). Individual scientists have a variety of goals, such as the desire for career advancement or prestige, which are not science's goals. Science's goals are the objectives that bring scientists together in pursuit of a common purpose, not the goals that many scientists, as individuals, happen to have. It is also useful to distinguish between interdisciplinary goals and discipline-specific ones. Interdisciplinary goals are goals that different disciplines share. For example, psychology and neurobiology both have the goal of understanding human behavior even though they have very different theories, methods, and concepts. Discipline-specific goals are goals that are unique to a particular discipline, such as paleontology's goal of discovering why the dinosaurs went extinct.

Do all of the different scientific disciplines share one common goal? Although different disciplines share many common goals, science has no single, defining goal. If it did, then it would be possible to define "science" as "the human activity that pursues this goal." One could then use this definition to distinguish between science, pseudo-science, and nonscience. The problem of defining science—also known as the demarcation problem—has proven to be very difficult, if not intractable. Decades of debates about this problem, among such noted philosophers and historians of science as Karl Popper, Thomas Kuhn, Paul Feyerabend, Imre Lakatos, and Philip Kitcher, have resulted in a stalemate. There is no universally agreed upon set of necessary and sufficient conditions for applying the term "science" that can satisfactorily distinguish between astronomy and astrology, creation science and evolutionary biology, medicine and homeopathy, or philosophy and psychology. The best one can hope to do is compile a list of characteristics that most of those human activities that we call science have in common (Kitcher 1993). To borrow a useful idea from Wittgenstein (1953), science is a family-resemblance concept.

One reason that science has no single, defining goal is that science is a highly complex and heterogeneous human activity, encompassing such fields as anthropology, economics, and psychology; as well as chemistry, physics, and biology; and also immunology, pharmacology, and pediatric oncology. These different scientific disciplines have diverse methods, theories, and concepts that cannot be unified into one common theory, methodology, or conceptual scheme (Dupré 1993). Science is more like a vast network of interacting states with common interests than like a regimented government. It resembles a business with decentralized management, not a business with centralized management.

Another reason that science has no single, defining goal is that science is politically, geographically, and culturally diverse. Science is practiced in countries all across the globe by people of different genders and of different racial, ethnic, and religious backgrounds, operating under different forms of government. Science has existed in ancient Greece; medieval Iraq; renaissance Italy; Enlightenment Prussia; Industrial England and America; Stalinist Russia; Nazi Germany; and in twenty-first-century Singapore, China, and India. Science is conducted at university labs, college labs, private labs, government labs, forensic labs, pathology labs, high school labs, and top-secret defense labs.

Since science is not a simple, homogeneous practice, it has many different discipline-specific and interdisciplinary goals. Science's interdisciplinary goals include epistemological goals, such as truth, knowledge, explanation, and understanding, and practical goals, such as prediction, control, power, and problem solving (Resnik 1998a; Kitcher 1993). These two types of goals correspond roughly to the distinction between basic and applied science: basic research pursues knowledge for its own sake, while applied research seeks knowledge with an eye toward practical applications. Although some scientific disciplines, such as quantum mechanics, tend to conduct basic research and pursue epistemological goals, and others, such as electrical engineering tend to conduct applied research and pursue practical goals, there is often no clear division between basic and applied science or between epistemological and practical goals. Basic science often has practical implications, and applied science builds on the theories and concepts from basic science. In some fields, such as biotechnology, genomics, electronics, and computer science, the historical division between basic and applied research makes little sense.

Before concluding this section, it is important to consider an objection to this account of science's goals. One might admit that science has many different goals and yet claim that all these different goals are subservient to one goal, the highest goal of science. Since science has historically been regarded as systematic knowledge, one could claim that knowledge is the highest goal of science. All other goals must serve this highest goal.

There are several problems with this objection. First, as noted above, it may run afoul of the demarcation problem: if knowledge were *the* goal of science, then it should be possible to define "science" as "the human activity that pursues knowledge." Many different disciplines that we would not call sciences pursue knowledge, such as philosophy, theology, law, and art history. Many pseudo-scientific disciplines, such as astrology and numerology, also claim to pursue knowledge. Since this definition does not adequately distinguish between science, nonscience, and pseudo-science, we have reasons to believe that knowledge is not the single, highest goal of science.

Second, if all the different goals of science served a highest goal, then it should be possible to prioritize these different goals when they conflict by appealing to the highest goal. One type of conflict that frequently occurs in science is the conflict between understanding and control. A theory, hypothesis or research program that helps us understand the world may not give us a great deal of control, and vice versa. For example, astrophysics helps us to understand the origins and fate of the universe, but it gives us very little control over the universe. Weapons research may provide us a great deal of control over human affairs, but it adds very little to our understanding of human affairs. If we are faced with a choice between funding astrophysics research and funding weapons research, how should we solve this dilemma? An appeal to some highest goal, such as the advancement of human knowledge, will not provide a useful solution, since both these research programs advance human knowledge in some way. These goals are—to use a term made famous by Thomas Kuhn—incommensurable (Kuhn 1970). We cannot compare them to a common metric, such as their ability to promote the advancement of human knowledge.

If science's goals are incommensurable, how can scientists (and nonscientists) resolve debates that involve questions about the goals of science? How should society decide between spending money on weapons research and spending money on astrophysics? One possible way out of this problem is to say conflicts that involve the goals of science should be resolved by appealing to higher social, moral, or political goals. This solution is consistent with Francis Bacon's (1626) idea that science should benefit mankind and improve the human condition. The images of corrupt, evil, or reckless scientists that one finds in the popular media elicit a visceral reaction because they tap into the public's fear that science does not benefit mankind and worsens the human condition.

There is a problem with this solution, however: not everyone agrees about the social, political, and moral goals science is supposed to serve. We do not agree what it means to "benefit mankind" or "improve the human condition." Some people may believe that having a strong military is society's most important goal, while others believe that having a strong military is not as important as other goals. For an even more problematic example, consider debates about federal funding for human ES cell research, discussed in chapter 1. Some people regard human ES cell research as morally acceptable and for the benefit of mankind, while others regard it as morally abhorrent and against the interests of mankind (Green 2001; President's Council on Bioethics 2002). Thus, this proposed solution to the problem of resolving debates about science's goals does not really solve anything at all; it simply shifts the problem to a different level.

Any "ultimate" solution to the problems related to conflicts among the goals of science must propose a way of resolving complex moral, political, and social controversies. This solution—if there is one—is well beyond the scope of this book. Gutmann and Thompson (1996) have developed an interesting and useful approach to the problem of resolving disagreements about basic values in a democracy. According to Gutmann and Thompson, when citizens in a democratic society disagree about moral, political, or social issues, they should engage in honest, open, fair, well-informed and reflective debates and give publicly defensible reasons for their views. The deliberative approach sets forth a procedure for resolving moral, social, and political disputes within a democracy but does not presuppose an over-arching theory of morality or social justice, such as utilitarianism, libertarianism, or natural law theory.

One might argue that academic science and industrial science have such different goals and norms, that any theory of norms and goals of science should treat these types of science separately. Indeed, there is considerable evidence to support the claim that academic science and industrial sciences operate according to different norms (Resnik 1998a; Krimsky 2003). In academic research, individual scientists are free to decide what problems they will work on, and when, where, and with whom to share data, materials, tools, and results. Academic scientists are supposed to be motivated by the search for knowledge or the desire to do good works for society, not by the pursuit of commercial interests. Academic scientists conduct basic research, not applied research. In industrial research, individual scientists do not have the freedom to decide what problems to pursue: the company's management decides its research goals and strategies. In industrial research, the company owns the data, materials, tools, and results, and the company decides when, where, how, and with whom to share them. In industrial research, scientists are not interested in knowledge or in doing good works for society; they are only interested in commercial interests. Finally, industrial science focuses on applied research rather than basic research.

Although there are clearly differences between academic and industrial science, these categories are not as different as one might think. First, although academic scientists enjoy a great deal of intellectual freedom, their freedom is limited by the availability of funding. Funding decisions are made by review committees at private companies, federal agencies, or private charities. Second, although academic researchers are free to share data, materials, tools, and results, they may choose not to do so in order protect their research from premature publication or to secure priority for intellectual credit or property. Third, many academic scientists are motivated not only by the search for knowledge and the goal of doing good deeds for society but also by their interest in personal or financial gain. Fourth, many academic scientists conduct applied research with practical and commercial value. Fifth, many industrial scientists have a great deal of freedom to choose research problems because some companies have decentralized research planning in order to promote creativity and innovation. Sixth, although private companies maintain tight control over data, materials, tools, and results, they are often willing to share these items, under certain conditions, and they frequently publish their results. Seventh, many private companies sponsor basic research as well as applied research. Eighth, many scientists have careers in academia and industry. Although most scientists begin their careers in academia, they may move to industry and then back to academia. Ninth, academic and industrial scientists now frequently collaborate on research projects. Thus, modern science does not fit the rigid dichotomy of academic versus industrial science.

2.3 Principles of Scientific Inquiry

Having developed an account of the goals of science, we can now describe some of the principles (or rules) that govern scientific inquiry. Since many different scholars and scientists have already written about these principles, this book will provide a quick synopsis of the rules and make some specific comments about a few of them. Before presenting this synopsis, some general comments are in order.

First, the principles should be understood as normative ideals that apply to scientific conduct not as descriptive generalizations of scientific conduct. It is important to keep this distinction in mind. The account of these principles presented here is philosophical and analytical, not empirical or experimental. Clearly, a high percentage of scientists may not obey these principles, but that is not the purpose of this inquiry. Even if many scientists do not adhere to these principles, one can still make the point that they *ought* to obey these rules. The principles have normative force even though people sometimes fail to obey them. A principle has normative force if people publicly endorse it and there is a system of rewards and punishments for enforcing it (Gibbard 1990). For example, most scientists would agree that honesty is an important principle of inquiry. One can accept this as a principle of inquiry without assuming that it accurately describes scientific behavior. Even if one knows that many scientists are dishonest, it would still be the case that they ought to be honest. Scientists routinely tout the importance of honesty and educate students about honesty. Science also has a system for enforcing honesty. For example, dishonest scientists can be sanctioned for scientific misconduct if their dishonesty involves fabrication of data, falsification of data, or plagiarism (Shamoo and Resnik 2003).

Although these principles are normative, not descriptive, empirical studies of scientific behavior are relevant to research and scholarship on the norms of science. Descriptive studies can provide scientists and science scholars with important information about whether rules are obeyed, why people disobey them, how people learn about them, and how they are enforced. However, these descriptive studies are not very useful in helping scientists and scholars to define, justify or interpret these rules. For example, consider scientific misconduct. One can approach this topic from an empirical or an analytical perspective. Empirical studies of misconduct attempt to estimate the incidence of misconduct and explain why it occurs (Steneck 2000). Analytical studies attempt to define or interpret "scientific misconduct" or give a justification for a particular definition or interpretation of "scientific misconduct" (Resnik 2003b). Both types of studies have their place in research and scholarship on research integrity, but they have different goals and adhere to different methodologies.

Second, the principles should be understood as guidelines (or rules of thumb) rather than absolute rules (Resnik 1998a). The reasons they should be understood as guidelines is that they may sometimes conflict, and when they do, scientists must choose between (or among) different principles (Kuhn 1977; Quine and Ullian 1978; Thagard 1992). For example, scientists should try to develop theories or hypotheses with explanatory power: good theories and hypotheses are able to unify disparate phenomena into one coherent, explanatory scheme. Scientists should also attempt to develop theories and hypotheses that are parsimonious: good theories and hypotheses provide a simple, economical, or elegant account of the phenomena (Rosenberg 2000). These two principles may conflict when scientists must choose between two theories that both fit the evidence equally well but differ with respect to their simplicity and explanatory power. In such a situation, scientists must choose between parsimony and explanatory power, since it will not be possible to adhere to both principles.

Third, normative principles govern the ethics of behavior as well as the ethics of belief (Lycan 1988). Some of the principles of research can be viewed as epistemological principles because they govern belief formation, while others can be understood as practical principles because they govern human action (the ethics of behavior). For example, the principle of parsimony is an epistemological rule that applies to the formation of scientific theories and hypotheses and the assessment of scientific evidence. Science's epistemological rules apply to research methods and techniques. A principle of credit allocation, on the other hand, is a practical rule that applies to recognizing individual contributions to scientific discoveries and inventions (Resnik 1998a). Science's practical principles apply to practices and traditions related to such areas of conduct as peer review, publication, authorship, teaching, and hiring and promotion.

Although it is useful to distinguish between epistemological and practical principles, some principles are both epistemological and practical. For example, honesty is an important practical principle that applies to the publication of scientific data and results, but it is also an epistemological principle that applies to the interpretation of scientific data and the assessment of scientific evidence. Three other principles—objectivity (avoiding bias), openness (sharing information), and carefulness (avoiding errors)—are also epistemological and practical, because they apply to belief formation as well as behavior.

Fourth, many of the different principles apply to one of more steps of the genesis of scientific knowledge, otherwise known as the scientific method. The following general framework applies to most scientific disciplines (based on Shamoo and Resnik, 2003; see also Giere 2004):

Step 1: Select a problem to study.
Step 2: Review the relevant literature.
Step 3: Propose hypotheses to solve the problem.
Step 4: Design tests or experiments to test hypotheses.

Step 5: Perform tests or experiments.
Step 6: Collect, record, and manage data.
Step 8: Analyze and interpret data.
Step 9: Submit data and results to peers for evaluation.
Step 10: Receive peer review of research.
Step 11: Publish/disseminate results.
Step 12: Allow scientific community to accept or reject results.

This framework is a simplified model of scientific research, of course. First, scientists rarely perform these steps sequentially; they often go back and forth between steps, making changes in their research plan along the way. For example, a scientist may change the experimental design after running into some problems performing an experiment, or a scientist might modify the hypothesis after collecting some data. Second, sometimes scientists perform the steps out of order. For example, in data mining, a scientist may analyze a large body of data and develop a hypothesis to explain patterns in the data. Third, sometimes scientists perform more than one step at the same time. For example, a scientist may pose a problem and a potential solution at the same time. Even though this framework is an oversimplification, it is still very useful in thinking about science. In subsequent chapters, I will refer to this framework in explaining how money can affect scientific inquiry.

We can now consider some principles (or norms) for scientific inquiry (based on Schrader-Frechette 1994; National Academy of Science 1995; Resnik 1998a; Shamoo and Resnik 2003; Haack 2003; Giere 2004).

2.4 Ethical Principles for Scientific Research

Honesty: *Be honest in all scientific communications. Do not fabricate, falsify, or misrepresent data or results; do not plagiarize.* Honesty is important not only in scientific publication but also in many other aspects of scientific conduct, such as applying for grants, hiring and promotion, mentoring, and negotiating contracts.

Carefulness: *Avoid careless errors, sloppiness, and negligence. Carefully and critically scrutinize your own work. Keep good records of all of your research activities. Use research methods and analytical tools appropriate to the topic under investigation.* As noted earlier, carefulness promotes the epistemological goal of avoiding erroneous belief. Since errors in belief can also have practical consequences, such as harms to human safety or health, carefulness also has an ethical dimension. Even though error is unavoidable in science and even essential to the development of knowledge, one might argue that scientists who do not take reasonable steps to prevent or correct errors are guilty of negligence (Thomsen and Resnik 1995; Mayo 1996).

Objectivity: *Eliminate personal, social, economic, and political biases from experimental design, testing, data analysis and interpretation, peer review, and publication. Seek to develop unbiased data, methods, and results.* This principle is important for achieving one of science's most important goals, the goal of producing objective knowledge. Many different science scholars have argued that science is inherently biased and cannot be objective. Chapter 3 will examine the objectivity of science in more depth. Objectivity also plays a role in the understanding of concerns about conflicts of interest (COIs) in research, since COIs can undermine objectivity. Chapter 5 will consider COIs in more depth.

Openness: *Share ideas, data, theories, tools, methods, and results. Be open to criticism, advice, and new ideas.* Openness is crucial for collaboration, creativity, accountability, and peer review. Secrecy can encourage isolation, corruption, and intellectual stagnation in research. Indeed, secrecy has been used in the history of science to protect illegal, incompetent, or unethical research. Although openness is important in science, it often conflicts with legitimate claims to protect the confidentiality of preliminary (unpublished) data or results, trade secrets, military secrets, peer review activities, and human subjects research records (Shamoo and Resnik 2003). Chapters 6 and 7 will explore conflicts between secrecy and openness in more depth.

Freedom: *Do not interfere with scientists' liberty to pursue new avenues of research or challenge existing ideas, theories, and assumptions. Support freedom of thought and discussion in the research environment.* Freedom is important for creativity, discovery and innovation in science. Scientists cannot develop new knowledge if they are not allowed to pursue novel areas of inquiry or challenge existing ideas. Scientists in modern, democratic societies enjoy a great deal of freedom, but this has not always been the case. The Catholic Church placed Galileo under arrest for defending his heliocentric astronomy. The Soviet Union censored and imprisoned scientists who challenged Lysenkoism, a doctrine that supported Marxist theories of human nature (Shamoo and Resnik 2003). Today, scientists in modern democracies rarely face the threat of censorship or punishment for their beliefs. However, since scientific research usually costs a great deal of money, the politics of research funding can result in economic threats to freedom of inquiry. Chapter 8 will explore these issues in more depth.

Credit: *Give credit where credit is due.* The principle is important for collaboration, trust, and accountability in research. Scientists who collaborate expect that they will receive proper credit for their contributions, such as authorship or acknowledgment. Since credit is important for career advancement and prestige, some of the bitterest disputes in science involve the allocation of credit. Proper credit allocation is also important for documenting the role of different scientists on a research project, in case there is a problem with the research, such as error or fraud (Resnik 1997).

Respect for intellectual property: *Honor patents, copyrights, collaboration agreements, and other forms of intellectual property. Do not use unpublished data, results, or ideas without permission.* This principle is also important for promoting collaboration and trust in research, since people will not be willing to collaborate if they are concerned that their ideas may be misappropriated. The principle is embodied in the intellectual property system, which protects intellectual property rights. This principle will be discussed in more depth in chapter 6.

Respect for colleagues and students: *Treat your colleagues and students fairly. Respect their rights and dignity. Do not discriminate against colleagues or students or exploit them.* This principle is important for developing a research environment and culture that embodies trust, collegiality, fairness and respect. It is also vital to mentoring, education, hiring, and promotion in science (Shamoo and Resnik 2003; Steneck 2004).

Respect for research subjects: *Treat human and animal research subjects with respect. Protect and promote animal and human welfare and do not violate the dignity or rights of human subjects.* This ethical principle in research is based on moral recognition of the importance of human rights and dignity and animal welfare. There are many different laws and regulations that pertain to human and animal research subjects (Shamoo and Resnik 2003; Steneck 2004). This book will not explore these laws and regulations in any depth, but chapter 5 will consider the question of whether financial interests can threaten the rights and welfare of human subjects in research.

Competence: *Maintain and enhance your competence and expertise through lifelong education. Promote competence in your profession and report incompetence.* This principle almost requires no explanation. This principle is important for maintaining professionalism and integrity in research, for advancing human knowledge, and for securing the public's trust. Incompetent researchers may undermine the public's trust in science and commit errors that threaten human life and the environment.

Confidentiality: *Protect confidential communications in research.* Scientists have access to a great deal of confidential information, such as grant proposals or papers submitted for publication, personnel records, patient records, and military or business secrets. As noted earlier, confidentiality sometimes conflicts with openness. For example, if a research team receives a request to share data prior to publication, they may to decide to refuse this request in order to protect their own career interests.

Legality: *Obey relevant laws and regulations.* Adhering to the law is important for many different reasons: for respecting colleagues, students, and research subjects; for protecting research institutions and organizations; for accountability; and for maintaining the public's trust. There are a variety of laws that pertain to research, including laws that protect human and animal subjects, laws related to laboratory safety, laws related to equal employment opportunity, and laws pertaining to research or financial misconduct. This book will not discuss these laws in any depth. For further discussion, see Steneck (2004).

Social responsibility: *Strive to benefit society and to prevent or avoid harm to society through research, public education, civic engagement, and advocacy.* This principle is important for securing and maintaining society's support for science. It also promotes accountability and trust. Viewed in this light, social responsibility embodies a contract that scientists have with society: scientists agree to help society in return for public support. Social responsibility also can be justified on moral grounds: scientists, like all citizens, have a moral duty to avoid harm and to benefit society (Schrader-Frechette 1994). Although most scientists recognize that they have social responsibilities, sometimes institutional pressures and personal interests, such as money or career advancement, can lead scientists to abdicate these responsibilities. Social responsibility may conflict with objectivity if scientists allow their political interests to skew their presentation of the facts when they act as advocates or engage in public debates.

Stewardship of resources: *Make fair and effective use of scientific resources. Do not destroy, abuse, or waste scientific resources.* This principle gives recognition to the fact that scientific resources are usually limited and sometimes fragile. Resources could include money, personnel, equipment, tissue samples, biological specimens, archeological sites, ecosystems, species, and human populations. Stewardship of resources is important for advancing scientific knowledge and for safeguarding moral values, such as respect for human beings, animals, or the environment.

2.5 Epistemological Principles for Scientific Research

Testability: *Propose theories and hypotheses that are testable.* Karl Popper (1959) defined science in terms of this principle: a scientific hypothesis or theory must be testable. Logical positivists, such as Ayer (1952), also championed this principle. Although many different philosophers have raised problems with testability in science, it still remains one the most important and influential principles in research (Mayo 1996; Giere 2004). Indeed, the U.S. Supreme Court cited the principle of testability in its ruling on the admissibility of scientific testimony in court (*Daubert v. Merrell Dow Pharmaceuticals* 1993).

Consistency: *Propose theories, models, or hypotheses that are internally consistent.* This principle makes good logical sense and has few detractors. Most scientists and philosophers of science regard internal consistency as a virtue (Giere 1988). Theories that are internally inconsistent are self-contradictory and impossible to test.

Coherence (conservatism): *Propose and accept theories or hypotheses that are consistent with other well-established scientific theories, laws, or facts.* Most philosophers of science accept this principle on the grounds that new theories or hypotheses should not cause major disruptions of well-established beliefs. If a hypotheses conflicts with a well-established principle, such as the conservation of mass-energy, one should consider rejecting the hypothesis before rejecting the well-established principle (Quine and Ullian 1978). One of the dangers of this principle is that it can lead to dogmatism. To avoid this problem, one must balance the desire for coherence against other epistemological norms.

Empirical support: *Proposes and accept theories and hypotheses that are well supported by evidence (data).* Science is first and foremost an empirical discipline: scientists use observations and data to support their theories and hypotheses (Van Fraassen 1980). Although some scientific disciplines, such as theoretical physics, are far removed from concrete observations, even these disciplines still depend on empirical support. While philosophers and scientists disagree about how to assess evidence, most agree that empirical support is essential to science (Giere 1988). Empirical support also has ethical implications insofar as scientists must trust that the data and results that they use have not been fabricated, falsified, or misrepresented manipulated. Dishonesty in science is contrary to the imperative to obtain empirical support.

Precision: *Propose theories and hypotheses that are precise and well defined.* Precision is important in science in order to test theories, models, and hypotheses, since a test of a poorly defined or imprecise statement will yield ambiguous or imprecise results. The computer programming maxim "Garbage in, garbage out" applies to almost any discipline, especially science. Precision, though important, is not an absolute requirement, since there are some limits to precision, especially in the social sciences. Many research methods used in social science, such as interviews, focus groups, and field observations, are not as precise as controlled experiments or quantitative measurements. Nevertheless, these methods can be useful in confirming hypotheses with explanatory and predictive power (Rosenberg 1988).

Parsimony: *Propose and accept theories and hypotheses that are simple, economical, or elegant.* English philosopher William Ockham developed the principle of parsimony (otherwise known as Ockham's Razor) during the Middle Ages. Other champions of parsimony include Isaac Newton and Albert Einstein. The main reason for following a principle of parsimony is not that the simpler theories are always true, since nature can be very complicated. The reason for seeking parsimony in science is pragmatic: simple theories are easier to understand, use, and test than complex theories. Although simple theories often turn out to be false, it is prudent to begin with a simple theory before developing a complex one (Lycan 1988).

Generality: *Propose, infer, and accept theories and hypotheses that are general in scope.* The reasons for following a principle of generality in science are also pragmatic: is it much easier to use a single, general theory for explanation and prediction than to use a patchwork of different theories. General theories also satisfy our desire to unify diverse phenomena into a coherent whole. General theories, like simple ones, often turn out to be false or at least riddled with exceptions. Even so, it is prudent to start with general principles before considering exceptions (Lycan 1988).

Novelty: *Propose, infer, and accept new theories and hypotheses; use new methods and techniques.* The importance of novelty is that it promotes creativity, discovery, and innovation in science. A new and interesting theory can launch different avenues of investigation and lead to the development of new concepts and vocabularies. Novelty can play an important role in counteracting conservatism in science (Kuhn 1970,

1977). Since novelty embodies newness in science, it also has important connections to freedom of inquiry, discussed earlier. Most scientific journals consider the novelty or originality of an article when deciding whether to publish it. Also, novelty is one of the criteria for patenting an invention.

2.6 Conclusion

To summarize this chapter, although people often think of science as value-free, it is not. Like other professions, science embodies different norms and goals. The norms of science help to advance the goals of science by promoting cooperation and trust among scientists and the public's support for science. They also help to ensure that scientific research is reliable, objective, and publicly accountable. Science has epistemological goals, such as knowledge, truth, and explanation, as well as practical ones, such as prediction, power and control. Science has epistemological norms, such as testability, empirical support, and simplicity, and practical norms, such as honesty, credit, and openness. Epistemological norms govern scientific methods and techniques, while practical norms apply to practices and traditions. The norms are not absolute rules but guidelines for the conduct of research. Sometimes the norms may conflict. When this happens, scientists must exercise their judgment to decide the best course of action to take. The next chapter will discuss science's most important norm, objectivity, in more detail.

THREE

SCIENTIFIC OBJECTIVITY

I never did give anyone hell. I just told them the truth, and they thought it was hell.

—President Harry S. Truman

3.1 Defining Objectivity

The previous chapter gave an overview of the norms and goals of science. This chapter will discuss science's most important norm, objectivity, in more detail. Objectivity is science's most fundamental value because many of the other norms of science, such as honesty, openness, carefulness, empirical adequacy, testability, and precision, are justified on the grounds that they promote objectivity in research. As we shall see in chapter 4, money's corrosive effect on objectivity poses a significant threat to the norms of science.

In chapter 2, objectivity was defined as a normative principle: one ought to eliminate biases from scientific research and develop unbiased methods and results. One might also view objectivity as a goal of science (Scheffler 1967). In either case, the claim is not that science *is* objective; the claim is only that it *ought to be* objective. This is an important distinction, since one might admit that science, qua human activity, often is not objective, yet still maintain that science ought to be objective. Likewise, one might believe that people often behave immorally, yet maintain that they should behave morally; or that the legal system is often unjust, yet still claim that it ought to be just. Objectivity, like justice, virtue, or piety, is a normative ideal (Scheffler 1967; Kitcher 2001; Haack 2003).

When people speak of science as objective, they sometimes take this to mean that science is value-free. This interpretation of the objectivity of science is mistaken and misleading since, as we saw in the previous chapter, science has its own epistemological and practical norms. A better way of interpreting the claim that "science is objective" might be to say that science is free from moral, political, or social values (Longino 1990; Kitcher 2001). However, given that science is practiced within larger societies or communities, which have their own values, the claim that science is free from moral, social or political values is also false or misleading. Moral, political, and social values affect scientific research in many different ways, ranging from the setting of government R&D priorities to laws dealing with research on human and animal subjects (Shamoo and Resnik 2003). So how should we define "scientific objectivity"?

Webster's Dictionary defines "objective" as "existing as an object or fact, independent of the mind; real . . . without bias or prejudice" (*Webster's* 1983). The *American Heritage Dictionary* defines "objective" as "having actual existence or reality . . . uninfluenced by emotions or personal prejudices" (*American Heritage Dictionary* 2001). *WordNet* defines "objective" as "undistorted by emotion or personal bias; based on observable phenomena . . . emphasizing or expressing things as perceived without distortion of personal feelings or interpretation" (*WordNet* 1997). These definitions emphasize two different senses of objectivity. In the first sense, something is objective if it is factual, true, or real; in the second sense, something is objective if it is not biased by personal feelings, emotions, or prejudices.

Some important government regulations on financial conflicts of interests in research equate "objective" with "unbiased." For example, the Public Health Service (PHS) has adopted guidelines for disclosure of financial interests titled "Objectivity in Research" (Public Health Service 1995). These guidelines apply to the NIH and the NSF. According to the guidelines, "This subpart promotes objectivity in research by establishing standards to ensure there is no reasonable expectation that the design, conduct, or reporting of research funded under PHS grants or cooperative agreements will be biased by any conflicting financial interest of an Investigator" (Public Health Service 1995). As one can see, this document attempts to promote objectivity by controlling the biasing effects of financial interests.

Philosophers have offered a variety of definitions of objectivity. According to Karl Popper, scientific statements are objective if they can be inter-subjectively tested (Popper 1959). Israel Scheffler contends that scientific statements are objective if they have been tested using impartial and independent criteria (Scheffler 1967). Philip Kitcher holds that the ideal of scientific objectivity is the idea that evidence and rules of deductive or inductive reasoning determine scientific theory choices, instead of personal, political, or cultural biases (Kitcher 2001). Sociologist and historian of science John Ziman equates objectivity with public agreement: scientific data or results are objective if similarly situated observers using the same techniques would produce the same data or results (Ziman 1968). The *Oxford Companion to Philosophy* defines an objective claim as a claim that is true or false, independent of what anyone thinks or feels (Honderich 1995).

Helen Longino has developed one of the most thoughtful accounts of objectivity in science and of the relationship between science and values. Longino distinguishes between two senses of scientific objectivity. In the first sense, to attribute objectivity to science is "to claim that the view provided by science is an accurate description of the facts of the natural world as they are." In the second sense, to attribute objectivity to science is "to claim that the view provided by science is one achieved by reliance nonarbitrary and nonsubjective criteria for developing, accepting and rejecting the hypotheses and theories that make up the view" (Longino 1990, 62).

This book will follow Longino's approach to objectivity and distinguish between two senses of the term. These two senses also roughly correspond to the two senses of "objectivity" found in dictionary definitions. The first sense is a metaphysical sense of "objective": a scientific statement, such as a hypothesis, observation, or theory, is objective if and only if it accurately describes a mind-independent reality. "Objectivity," in this sense, roughly means "factual, true, or real." A scientific method is objective, in this sense, if yields data or observations that accurately describe a mind-independent reality. *Realism* is the view that science describes a mind-independent reality (Giere 1988).

The second sense of "objectivity" is an epistemological one: a scientific statement is objective if and only if it is unbiased. For a statement to be unbiased, it must be based on reasoning and evidence, not on personal prejudices, feelings, opinions, economic interests, or moral, social, or political values (Resnik 2000a). For example, one might argue that randomized, controlled trials are an unbiased type of research method in clinical research because they tend to yield unbiased results. A scientific method is objective, in this sense, if it tends to yield statements that are unbiased. *Rationalism* is the view that science is based on reasoning and evidence (Newton-Smith 1981; Giere 1988).

As noted earlier, one may distinguish between descriptive and normative senses of objectivity. If we apply this distinction to the distinction between realism and rationalism, then we can distinguish between four different types of scientific objectivity: (1) *descriptive realism*, which holds that science accurately describes a mind-independent reality; (2) *normative realism*, which holds that science ought to describe a mind-independent reality; (3) *descriptive rationalism*, which holds that science is unbiased; and (4) *normative rationalism*, which holds that science ought to be unbiased.

Most scientists (and many philosophers of science) accept descriptive and normative versions of realism and rationalism. What are the arguments for these positions? Most arguments for objectivity begin with an account of scientific knowledge or the relationship between science and the world (science's metaphysics). The argument presented here will be different from these. It will start with an account of science's role in society and argue that science ought to be objective, before considering whether science is objective. The rationale for this argumentative strategy is that it forces us to think clearly about our ideals before deciding whether they can be realized. Of course, it may turn out that these ideals are illusions, but hopefully, this is not the case (Kitcher 2001).

3.2 Normative Rationalism

Let us begin by developing an argument for the claim that science ought to be unbiased (normative rationalism). The argument for this thesis is a political one: its conclusion is that democratic societies need science to inform public policy. The first premise in the argument is the assumption that the type of society in which science is conducted has a democratic form of government.[1] Although democracy is not a perfect form of government, it is preferable to the alternatives, such as fascism, oligarchy, monarchy, theocracy, or anarchy. The next premise is another assumption: the society in which science is conducted is pluralistic; that is, people in the society do not all share a common conception of the good, a common religious doctrine, or a common political ideology (Rawls 1993). Given these two assumptions, how can a society reach some reasonable agreement about laws and policies that apply to controversial moral, political, and social issues without sacrificing democratic principles or deteriorating into anarchy? To remain stable, democratic societies need some beliefs and methods for settling controversial, public issues. Science provides these methods and beliefs.

As noted in chapter 2, Gutmann and Thompson (1996) have developed an approach to this problem known as deliberative democracy. Citizens in democratic, pluralistic societies should decide controversial matters of public policy by engaging in honest, open, fair, well-informed, and reflective debates and by giving publicly defensible reasons for their views. Science can play an important role in this process by providing different parties in these debates with beliefs and methods that are perceived as neutral with respect to the very moral, political, economic, cultural, or social issues that are at stake. The participants need a third party that they regard as an unbiased purveyor of knowledge and information, and science can fulfill this important function. To borrow a phrasing from Voltaire, if science did not exist, it would be necessary to invent it.

To understand how science may serve as a neutral party, consider two current public health controversies. The debate about smoking in public pits personal freedom (the right to smoke) against social good (public health). Science has provided some useful information in this debate about the effects of second-hand smoke (Environmental Protection Agency 1994). This information has not completely resolved the debate, but it has helped society to form a consensus around some common rules. It is likely that such progress would not have been made in this controversy without the benefit of some information that the parties regarded as unbiased. Without this information, the debate would have remained at the level of personal freedom versus social good. Since both sides agree that personal freedom can be restricted (to a certain extent) to prevent harm to others, information about the effects of second-hand smoke can help bring the two sides toward some resolution of the debate. Also consider the debate about approving new drugs for human use. This type of debate pits drug manufacturers against advocates for patient safety and involves difficult trade-offs between safety, effectiveness, and access to medications (Angell 2004). To resolve a debate about approving a new drug, one needs to have knowledge about the new drug that is perceived to be neutral and impartial with respect to the various interests and values at stake in this debate. Since opposing sides agree (to a certain extent) that new drugs should be safe and effective, information about drug safety and efficacy can help to resolve a debate about approving the drug.

Science's role as a neutral party is most apparent when scientists are asked to give expert testimony in the courtroom or for governmental proceedings. When scientists serve as experts, they have an ethical obligation to be objective, namely, unbiased, fair, and impartial (Shamoo and Resnik 2003). The conflict of interest problems in the FDA's approval process provide a perfect illustration of why it is important for the public to view experts as neutral and impartial. One way to discredit the testimony of an expert witness who testifies before a court is to demonstrate that the witness has particular financial interests that favor one of the parties. Experts typically earn several thousand dollars for giving testimony in a particular case, but they may earn ten times that much (J. Murphy 2000). It is fair and proper, in a court of law, to ask an expert witness if he has any relationship to one of the parties or if he is being paid by one of the parties (J. Murphy 2000). Although expert witnesses take an oath to tell the truth, they might still break this oath or stretch, bend, or distort the truth.[2] In response to the problem of bias among expert witnesses, several professional associations have adopted ethical rules for witnesses giving testimony and for attorneys procuring expert testimony (J. Murphy 2000). Critics of the thesis that science ought to be unbiased might conclude that these problems with expert witnesses support their view that science is biased and partial. However, we should infer something different: the fact that we want to eliminate potential biases among expert witnesses (and science advisors to the FDA) shows we value scientific objectivity.

The argument to this point has attempted to prove that scientists ought to strive to be perceived as rational (or unbiased) by the public, since people need to have as a source of knowledge and information that they regard as unbiased (or neutral or impartial). But scientists ought to do more than try to appear to be unbiased; they should attempt to be unbiased. Two arguments support this claim. First, it is much easier and simpler to try to be something instead of trying to appear to be something. Consider someone who wants to appear to be honest but does not want to actually be honest. He puts on a good show for the public but is secretly dishonest. This strategy would consume a great deal of energy and would ultimately backfire. The person who fakes honesty must keep up with many different lies and will eventually be discovered as a fraud. The same argument applies to science. Scientists, who pretend to be unbiased but are secretly biased, will expend a great deal of energy trying to present an unbiased image and will eventually be discovered. The best way to appear to be unbiased is simply to try to be unbiased.

Second, the public would lose all of its trust in science if people ever found out that scientists were only striving for the appearance of objectivity. This would be like finding out that someone who has been claiming to tell the truth has only been trying to tell a story that people would believe is the truth. If this liar is ever found out, no one will ever believe anything else he says. To avoid these problems, scientists must strive to be objective, not just to appear to be objective.

3.3 Objections to Normative Rationalism

Aside from challenging its first two assumptions, how might one critique this argument for the claim that science ought to be unbiased? First, one could argue that science does not need to be unbiased or appear to be unbiased. One might argue, for example, that science should not be a neutral party in social, political, moral, or economic debates, but that it should be an active participant. Every human activity or profession embodies a set of moral, social, economic, cultural, or political values, including science. Scientists should accept theories and hypotheses based on their ability to advance particular moral, political, social, or economic goals, rather than on their ability to account for the evidence.[3] At bottom, all debates in the world are political debates. Scientists should recognize and accept this fact and not hide behind an imaginary cloak of rationality.

This objection has some serious flaws. First, in chapter 2 it was argued that the ideal of objectivity is part of science's normative structure. If this is the case, then scientists should try to accept theories and hypotheses based on arguments and evidence. To reject the commitment to rationality would be, in effect, to reject science.

Second, if people regard science as no more than another interest group with its own set of social, political, economic, cultural or moral values, then why do people allow scientists to give expert testimony and why do they criticize scientists who advocate for a particular moral, social, political, cultural or economic viewpoint? If science is no different from a religion, a political party, or a social movement, then the public should treat scientists the same as they treat people who represent particular religious, moral, social, or economic viewpoints. But people do not treat scientists this way when they ask scientists to provide expert testimony or public information; they treat scientists as neutral and impartial.

The second objection to the claim that scientists should be unbiased is not dismissed as easily as the first. According to the second objection, scientists should not attempt to be unbiased because they cannot be unbiased. Science is irrational. Scientists do not—and cannot—make decisions based on arguments and evidence. They make all their decisions based on various moral, political, social, cultural, or economic biases. Rationality is, at best, a delusion; at worst, it is a fraud perpetrated on the public. Since one does not have an obligation to do something that one cannot do (i.e., "ought implies can"), scientists do not have an obligation to be unbiased.

Since the first publication of Thomas Kuhn's *The Structure of Scientific Revolutions* (1962) over four decades ago, many different philosophers, sociologists, and historians of science have attacked science's objectivity by claiming that science is not rational (Newton-Smith 1981; Klee 1997). Kuhn made a number of different remarks in his influential book that encouraged people to view science as irrational. Kuhn attacked the cumulative model of scientific progress, which holds that science is a steady march to the truth based on careful reasoning and evidence. According to Kuhn, scientific change often takes place through revolutions and there is no march toward the truth. Kuhn's model of change held that science cycles through two different periods, which he called normal science and revolutionary science. A new field emerges from pre-scientific instability when the practitioners accept a common paradigm. Kuhn used the word "paradigm" in at least two senses: according to the first sense, a paradigm is a disciplinary matrix; according to the second, it is a shared exemplar. A disciplinary matrix includes assumptions, concepts, theories, methods, evidence regarded as relevant, traditions, tools, and techniques. A shared exemplar is a widely accepted body of work that shows how to solve research problems within a disciplinary matrix. Scientists who accept a common paradigm are practicing what Kuhn called "normal science." During a period of normal science, scientists attempt to articulate, apply the paradigm to various phenomena. For example, following the publication of Newton's *Principia*, the Newtonian paradigm dominated physics. The *Principia* served as the shared exemplar for this paradigm, and Newtonians accepted common assumptions, beliefs, concepts, methods, and so forth.

Periods of normal science do not last forever. Eventually, some scientists begin to become concerned with phenomena that the paradigm cannot explain, otherwise known as anomalies. Soon a crisis emerges, as scientists propose new theories and develop new concepts and methods to deal with the anomalies. As a result of this discontent, a scientific revolution soon takes place. Scientists abandon the old paradigm in favor of a new one, and a new period of normal science begins (Kuhn 1970). For example, during the early 1900s, physicists became concerned about Newtonian physics' inability to adequately account for electromagnetic radiation. Physicists developed new theories to explain electromagnetic phenomena, and quantum mechanics emerged as a new paradigm in physics.

According to Kuhn, every theory must be understood within its associated paradigm, which defines the meaning of the terms used in the theory. Different theories may use the same words but give them very different meanings. For example, "mass," "space," and "time" have very different meanings in Newtonian physics and in post-Newtonian physics (quantum mechanics and relativity theory). Since different theories use terms with different meanings and are judged according to different evidence-bases, one cannot compare two scientific theories according to a common metric. Theories are incommensurable. Since theories are incommensurable, scientists do not change theories as a result of reasoning and argumentation. "The superiority of one theory over another is something that cannot be proved in a debate. . . . [E]ach party must try, by persuasion, to convert the other" (Kuhn 1970, 198). Kuhn also held that paradigms are incommensurable. Kuhn compared a paradigm to shift to a gestalt switch. According to Kuhn, "When the normal-scientific tradition changes, the scientist's perception of his environment must be re-educated—in some familiar situations he must learn to see a new gestalt" (Kuhn 1970, 112). The incommensurablity of theories and paradigms opens the door to nonrational explanations of scientific change, such as social, political, economic, cultural or psychological explanations (Newton-Smith 1981).

Many people who read Kuhn interpreted him as saying that science is not rational, especially since he used phrases like "gestalt" switch and "conversion" to describe theory change in science. In the second edition of *The Structure of Scientific Revolutions* (1970), Kuhn wrote a postscript in which he tried to clarify his view about scientific rationality. He also expanded these views in *The Essential Tension* (1977). Kuhn claimed that he did not reject scientific rationality outright but thought that it needed to be rethought in light of his observations about the history of scientific change. He suggested that scientific rationality is based, in part, on shared commitments to values, such as simplicity, precision, testability, honesty, generality, and the like. These values play a key role in the decision to accept one theory and reject another one.

Adopting a new paradigm is more than putting on a new pair of lenses to look at the world; a paradigm shift literally creates a new world. According to Kuhn, after a revolution, scientists are living in a different world (Kuhn 1970). The world has changed because what the world is depends on the concepts, theories, and beliefs of people who observe the world. Although science makes progress—it becomes better at solving problems—it does not make progress toward a true description of a mind-independent world (Kuhn 1970, 206). Scientific statements do not describe or represent a mind-independent world or the "facts."

Although Kuhn did not say that science is socially constructed, many of the critics of scientific objectivity, who have taken inspiration from Kuhn, regard themselves as (or have been labeled as) social constructivists (Hacking 2001). Some of these writers include Pickering (1992), Bloor (1991), Latour and Woolgar (1986), and Knorr-Cetina (1981). While there are many different versions of social constructivism, they all share some common assumptions. First, they regard science as a type of practice or culture like any other practice and culture. One could study the beliefs and practices of physicists as one could study the beliefs and practices of Roman Catholics, Zulu warriors, or the Australian Bushmen. There is nothing special or unique about the way scientists form there beliefs, nor is there anything that distinguishes a modern, scientific explanation of the rising and setting of the sun from a pre-modern, unscientific explanation. Knowledge is relative to a particular society or culture (Bloor 1991). Second, the constructivists argue that science is not rational. Scientists do not change their beliefs, hypotheses, or theories as a result of reasoning and evidence; political, social, economic, and psychological factors explain epistemological change in science. Third, they claim that science does not accurately represent a mind-independent reality. Science is socially constructed, invented, fabricated, or created. Scientists do not make progress toward the truth, nor do they make discoveries about a pre-existing, independent world. Fourth, science is not a neutral or impartial purveyor of facts; science embodies political ideologies or social agendas, such as racism or sexism (Harding 1986).

There is not sufficient space in this book to explore in great depth these debates about the rationality of science. But we do not need to definitively settle these questions in order to defend the view that scientists should try to be unbiased. To understand objections to the thesis that scientists ought to be unbiased, we should distinguish between two different critiques of this view: an empirical critique and a conceptual one. According to the empirical critique, scientists should not strive to be rational because science is not, as it turns out, rational. Science could be rational, but it just so happens that it is not rational, due to human weakness and fallibility. According to the conceptual attack, scientists should not strive to be rational because science cannot be rational; scientific rationality is a logical or conceptual impossibility. Talking about scientific rationality makes as much sense as talking about round squares.

To support the view that scientists ought to be unbiased, one does not need to prove that science is unbiased. Science may be biased to some degree, perhaps even to a high degree. The question of exactly how different biases affect science is an important empirical and conceptual problem that is best settled by additional studies in the history, philosophy, sociology, and the psychology of science (Giere 1988). The query in this chapter does not address empirical questions about how scientists behave. Rather, it is concerned with how scientists ought to behave. Empirical evidence relating to political, social, psychological, economic and cultural influences on science does not demonstrate that science should not seek objectivity. For comparison, one might hold, on the one hand, that people are sinners, and, on the other hand, that people ought to avoid sin. One might hold, on the one hand, that no human society is perfectly just, but also hold, on the other hand, that people should strive for justice.

To defeat the thesis that science should aim for objectivity, one must argue that it is impossible for science to be unbiased. If it is impossible to be unbiased, then one should not strive to be unbiased. How might one argue that science cannot be unbiased? One could argue that scientists never respond to arguments and evidence. Even when it appears that scientists accept a hypothesis based on the weight of the evidence in its favor, they are really accepting the hypothesis as a result of psychological, sociological, economic, cultural, or political factors. Scientists cannot even choose to be unbiased, because they will simply choose to favor one bias over another bias. Striving to be unbiased is like trying to develop a perpetual motion machine; it is a conceptually impossible and futile quest. It is better to give up the quest than to be seduced by it.

The reply to this argument is that it is possible for science to strive to be unbiased because human beings conduct scientific research, and human beings are capable of striving for objectivity. To strive for objectivity, one must be able to decide questions not based on political, social, cultural, economic, or other influences, but based on well-reasoned arguments and clear evidence. One must be able to accept a belief or theory not because it promotes one's political, social, economic, or other interests, but because it is well supported by the arguments and evidence. Although human beings are highly biased creatures, they are still capable of making choices based on arguments and evidence. If it were not possible for human beings to respond to arguments and evidence, then there is no point in engaging in any activity where one presents or contemplates arguments and evidence. To even understand this book, one must assume that it is possible for human beings to respond to arguments and evidence. If human beings can respond to arguments and evidence, then can't scientists also respond to arguments and evidence? Is there something unique about scientific research that blinds people to reasoning and evidence, while ordinary human life does not have this effect? On the contrary, one would think the opposite would be true: people are more likely to be affected by reasoning and evidence when they engage in scientific activity than when they engage in other human activities, such as religion, politics, business, and so on.

3.4 Normative Realism

So far we have argued that scientists ought to be unbiased. This conclusion serves as a basis for the thesis that scientists should seek to describe a mind-independent reality. If we start from the premise that scientists ought to be unbiased, we can ask ourselves, what is the best way to achieve the goal? Would it be more effective to try to develop an accurate representation of a mind-independent world, or would it be more effective to develop a mind-dependent construction? Would there even be a detectable difference between the two? Let's consider a situation where there might be a difference and a preference, such as developing a map. Map-making is a useful metaphor for discussing realism in science, since many different scientific disciplines create maps (Kitcher 2001). Scientists have developed maps of many different physical things, such as chromosomes, cells, the bottom of the sea, the surface of the moon, and the galaxy.

There are two very different ways of making a map of a city, a mind-independent (or realist) strategy and a mind-dependent (or constructivist) one. The realist strategy develops a map of a city by treating the different streets and landmarks as mind-independent objects related to other objects in space. This strategy requires the mapmaker to observe the city and to develop a model of the city (the map) that represents the city's spatial relationships. For example, if street A is twice as long as street B in the city, then it should be twice as long in a map of the city. The constructivist strategy develops a map of the city by treating the different streets and landmarks as mind-dependent objects related to each other in a collection of stories and verbal reports. This strategy collects information about the city from people who claim that they have lived in the city, been to the city, or heard about the city from other people. This strategy requires the mapmaker to try to mesh all of these different accounts of the city into a single, logically coherent representation of the city.

Suppose that someone challenges the maps, claiming that they are biased. He claims that the spatial relationships in the maps result from economic, political, and social biases because, on the map, the streets where the rich people live are much wider than the streets where the poor people live. The person who pursues the realist strategy has a straightforward way of determining whether the map is biased: he can compare places on the map to places in the city to ascertain whether the map preserves roughly the same proportions as found in the city. If it does, then the map accurately represents the city and he can argue that the map is not biased. If the map does not accurately represent the city, then he can make changes in the map so that it will be less biased. Eventually, by making corrections in his map, he can develop a more accurate (less biased) map. The person who pursues the constructivist strategy will find it difficult or impossible to prove that his map is unbiased. If he checks the map against his collection of verbal accounts, a critic can always argue that he has not gathered an impartial or unbiased sample of verbal reports or that the reports themselves are biased. Since he cannot check his map against a mind-independent reality, this mapmaker will always be susceptible to the charge that he is biased. Indeed, even he himself will not be able to be sure that his map is not biased. The best he can hope to do is develop a map that everyone agrees is not biased. But this agreement, by itself, does not prove that his map is unbiased, nor does less disagreement prove that his map is less biased.

Granted, not all scientists produce maps, but most scientists create models (Giere 1988). The conclusions drawn in the last two paragraphs also apply to developing scientific models. One could ask: what is the best strategy for building an unbiased model—the realist strategy or the constructivist one? The realist's strategy appears to be a better strategy for developing an unbiased map or model. (As an aside, the constructivist's strategy may be better at performing another task, such as interpreting a legal document, developing a screenplay from a book, or writing a biography.) Thus, from the claim that science ought to be unbiased one can draw the conclusion that science ought to try to develop accurate representations of mind-independent objects. This does not prove, of course, that scientists actually can have completely accurate representations of mind-independent objects, only that they ought to strive for such representations.

3.5 Objections to Normative Realism

If only it were so simple. Critics of objectivity in science will protest that the map-making metaphor does not apply to scientific research and even begs the very question at hand. One again, we can distinguish between two different critiques of the realist's strategy: an empirical critique and a conceptual one. According to the empirical critique, science does not describe a mind-independent world. To prove this claim, a critic would appeal to empirical evidence from the history, sociology, and psychology of science to show that scientists have failed to describe a mind-independent reality. According to the conceptual critique, it is impossible for science to describe a mind-independent reality. To prove this claim, a critic would need to develop a philosophical argument against realism. Let's consider the empirical attack first.

To evaluate both of these critiques, it is important that we set aside questions about whether science's theoretical statements accurately describe reality. According to some philosophers of science known as "anti-realists," science cannot provide evidence for the truth of its theoretical statements because these statements describe objects, properties, processes, and events that are not directly observable. For instance, people cannot directly observe atoms, molecules, DNA, chromosomes, or cells, but they can directly observe the reading on a thermometer, the number of eggs that a turtle lays, or the sound that an owl makes. A scientific "anti-realist" would deny that we have proof for the existence of atoms, molecules, or DNA, although he or she would grant that we have evidence for the existence of thermometers, turtle eggs, or owls, since we can directly observe these things (Rosenberg 2000, Klee 1997).[4]

While this is an important and interesting debate in the philosophy of science, the social constructivist attacks on science go much deeper than the "anti-realists" attacks on science's theoretical statements. The person who denies descriptive realism claims that science does not accurately describe a mind-independent world at all. Science's theoretical and observational statements are socially constructed. Even statements such as "Water freezes at 0° Celsius," "Human beings have a four-chambered heart," and "Oak trees shed their leaves in autumn" do not describe a mind-independent reality. This claim conflicts with scientists' own understanding of what they are doing as well as with our commonsense ideas about science (Fine 1996; Haack 2003).

One of the most influential empirical arguments against descriptive realism takes its inspiration from Kuhn's reading of the history of science. The argument goes like this: The history of science is a giant scrap heap of mistaken theories, beliefs, concepts, hypotheses, and assumptions. Quantum mechanics overturned Newton's mechanics, which replaced Galileo's mechanics, which overturned Aristotle's mechanics. One can reasonably infer that some day another theory will come along and show that quantum mechanics is mistaken. At one time, scientists believed in phlogiston, the ether, spontaneous generation, vital forces, and biologically distinct races, but they no longer believe in these things. History shows us that today's best theories are probably false, and that many of the things scientists regard as real will turn out to be illusions. Thus, science does not describe a mind-independent reality at all. It is simply a succession of different theories, hypotheses, beliefs, assumptions, and concepts. Science may become better at solving problems, but it does not provide us with the truth, or even make progress toward the truth (Kuhn 1970).

Advocates of descriptive realism read history differently. While they admit that scientists have held many erroneous beliefs, theories, and hypotheses throughout history, they claim that science has been a series of practical and epistemological successes, not a series of failures. Science has been incredibly successful at predicting and explaining many different phenomena and in helping people obtain mastery and control over the world. It is highly unlikely that this success is due only to social, economic, or political circumstances or mere luck. The best explanation of science's success is that science accurately describes a mind-independent reality and is making progress toward a better description of that reality (Boyd 1984; Kitcher 1993).

Science's successes are practical and epistemological. From a practical point of view, science has been very successful at enabling people to achieve practical goals, such as sending men to the moon or developing a polio vaccine. How could we possibly achieve these complex, technical goals without scientific knowledge that accurately describes mind-independent objects? From an epistemological viewpoint, one can point to a vast body of well-confirmed statements and laws in many different disciplines that have not changed in many years. The science of human anatomy has not changed much since the time of Leonardo da Vinci. The anatomists' descriptions and illustrations of the basic organs, muscles, bones, and tissues of the human body accurately describe the human body. Other disciplines that have experienced a great amount of stability include parts of chemistry, geology, biology, and astronomy. In the sciences where a great deal of disagreement continues, such as particle physics or astrophysics, one can see a steady progression toward the truth. Even though quantum mechanics has replaced Newtonian mechanics, Newtonian mechanics makes very accurate predictions for situations where the objects are larger than atoms and do not move at speeds approaching the speed of light. Newton's theory is a very good approximation, and it is a better approximation than Galileo's theory (Kitcher 1993).

The deeper challenges to normative realism come not from sociology or history, but from philosophy. The philosophical argument against normative realism is that it is impossible for scientific statements to accurately describe a mind-independent world because there is no mind-independent reality, or, if there is one, we cannot know anything about it. The world as we know it is a mental (or social) construction. If it is impossible for scientific statements to accurately describe a mind-independent world, then scientists do not have an obligation to develop statements that describe such a world, since one cannot have an obligation to do something that one cannot do.

These philosophical challenges to realism reflect disputes about the nature of reality that trace their history all the way to Plato and Aristotle's debate about theory of forms. This book will not review the entire history of metaphysics here, but it will mention some key points in the debate from the scientific revolution onward, since this will help us to understand challenges to the idea the science should be unbiased and the social constructivist perspective on metaphysics.

(Philosophers, please bear with this very oversimplified version of the history of philosophy.)

After the scientific revolution (circa 1450–1650), a philosophical debate emerged concerning the limits of human knowledge, which pitted realists against idealists. The realists argued that human beings could obtain knowledge concerning a mind-independent reality, while the idealists claimed the human beings could only obtain knowledge of a mind-dependent world. The English philosopher John Locke, who wanted to help lay the conceptual foundations for the rapidly advancing sciences of mechanics, astronomy, optics, chemistry, anatomy, and medicine, argued that there are mind-independent objects, known as substances, that have different properties, which he called primary, secondary, and tertiary qualities. Primary qualities are properties that are really in physical objects, such as mass, size, shape, and motion; secondary qualities are properties that are not in physical objects but are produced in the minds of observers by physical objects, such as color, sound, texture, and so on; and tertiary qualities are properties that exist in physical objects that allow them to cause changes in other physical objects, such as the power of a hammer to break glass. Locke held that science can study the objective properties of objects (primary and tertiary qualities) but not the subjective properties of objects (Locke 1690). Locke was also an empiricist: he believed that all scientific knowledge must be based on evidence from the senses. Thus, Locke's philosophy of nature introduces a distinction between objectivity and subjectivity and treats science as objective. Locke believed that science can study substances and provide us with knowledge about their primary and tertiary qualities, which are objective.

Bishop George Berkeley, a seventeenth-century English philosopher and theologian, accepted Locke's empiricism but attacked his metaphysics. According to Berkeley, reality is not mind-independent; only minds and ideas are real. Berkeley's argument for this view begins with a commitment to empiricism: all of our knowledge must be based on information from our senses. Sensory information produces sensory ideas, such as color, texture, sound, and the like, as well as ideas formed by attending to operations of our own mind or from memory, such as ideas of time, motion, or causation. According to Berkeley, we can only compare our ideas to other ideas; we cannot compare our ideas to mind-independent objects. For example, suppose we ask whether an apple and a sweater are the same color of red. What allows us to judge that the apple is redder than the sweater is that we can compare our two ideas of redness to each other, not that we can compare our ideas to any mind-independent object. Even when we use language to talk about the so-called primary qualities of physical objects, we are still only referring to ideas, which have been generated by our senses. Thus, when we say, "An apple is smaller than the sweater," we are only talking about relationships among our ideas, such as "apple," "smallness," and "sweater." We are not talking about something "out there" in a mind-independent world. Thus, Berkeley rejected Locke's distinctions between different types of qualities and held that all qualities exist only in minds. For Berkeley, *esse est percepi*: to be is to be perceived (Berkeley 1710).

The eighteenth-century Scottish Philosopher David Hume followed in Locke's and Berkeley's empiricist tradition and took idealism to its logical conclusion. Hume went much further than Berkeley in his skepticism about a mind-independent reality. Hume rejected the belief in substances, primary qualities, causal necessity, God, and even the self (or soul) on the grounds that these ideas are about things that are beyond the realm of human experience. According to Hume, if we cannot trace belief in an object, property, or process back to its source in sensory perception, then that belief is unjustified and should be rejected as idle metaphysics. Although Hume developed powerful philosophical arguments against various metaphysical beliefs on the grounds that they have no basis in experience, he did acknowledge that we have very good practical reasons for accepting some of these beliefs. For example, one cannot walk very far without believing in mind-independent objects. Although Hume acknowledged that it is legitimate to believe in mind-independent objects, he did not think that we can have knowledge of those objects. Human knowledge is limited to human experience (Hume 1748).

The eighteenth-century German Philosopher Immanuel Kant responded to the positions developed by Locke, Berkeley, and Hume. Kant developed a position known as transcendental idealism, which accepts Berkeley's claim that our knowledge is mind-dependent but also accepts Locke's view that there is a mind-independent world "out there." He accepted Hume's claim that all knowledge must be based on experience, but he argued that the human mind has concepts that provide a structure to experience. According to Kant, a system of fourteen fundamental concepts provides the foundation for all of our scientific knowledge. Kant held that these concepts are a priori, which means that they are not based on or derived from experience. Kant's system of concepts consists of the concepts of time and space as well as twelve concepts known as the categories, which include the concepts of reality, negation, unity, plurality, substance, causation, possibility, and existence. We acquire scientific knowledge, according to Kant, when we apply these fourteen concepts to the raw information that we receive from our senses. Indeed, we do not perceive anything at all unless we use these basic concepts to perceive it. There is no such thing as raw, unprocessed sensory experience; all sensory perception is perception by means of concepts. The knowledge that we obtain from our senses and our fundamental concepts constitutes knowledge of things as they appear to us (or the phenomena). This world is objective in the sense that it is a world of objects, but it is not a world of mind-independent objects. Thus, for Kant, there is a world "out there" but we cannot know it; we can only have knowledge of a world that is mind-dependent (Kant 1787). Even so, that knowledge can be "objective," and "real" in Kant's special senses of these words, because it can be knowledge about real objects in space and time. However, the concepts of space, time, objects, and reality are simply are ways of representing reality to ourselves.

Even though Kant did not believe that we can have knowledge of a mind-independent world, he held that we must posit the existence of a world of mind-independent objects, such as the soul, freedom, and God, in order to justify morality and religion. We can have no scientific knowledge of these mind-independent objects, known as things-in-themselves (or noumena), but we must believe in them nonetheless.[5] If we did not believe in the soul and freedom, for example, then we would regard people as like machines, and the concept of moral obligation would been pointless. If we did not believe in God, morality would seem to be futile.

In the twentieth century, pragmatists developed a different approach to questions about belief in a mind-independent world. Pragmatists generally believe that metaphysical questions are best settled by determining what works well in practice. William James updated the practical argument for belief in God by claiming that it is rational to believe in God if this belief proves to be useful and productive in one's life (James 1898). John Dewey believed that a philosophy must be tested by the way it works out in practice (Dewey 1910). Long before Kuhn reflected on scientific progress, Charles Peirce had recognized that science was fallible and would continue to make mistakes. Peirce, however, did not reject the idea of scientific progress:

> The followers of science are animated by a cheerful hope that the process of investigation, if only pushed far enough, will give one

> certain solution to each question to which they apply it.... This great hope is embodied in the conception of truth and reality. The opinion which is fated to be ultimately agreed upon by all who investigate, is what we mean by the truth, and the object represented in this opinion is the real.... Reality is independent, not necessarily of thought in general, but only of what you or I or any number of men may think about it. (1940, 38)

W.V. Quine, though not a follower of James, Dewey, or Peirce, studied under one of their contemporaries, Alfred North Whitehead. Quine synthesized pragmatism and logical positivism. The logical positivists attempted to use logical and linguistic analysis to purge science of metaphysics. They held that statements are meaningful only if they assert a claim about a possible observation or they express a logical or mathematical truth. The positivists sought to rid science of meaningless statements. To do this, they attempted to develop a new linguistic framework (or language) that would translate all scientific statements into statements about potential observations or statements about mathematics or logic. In this way, science could realize Hume's dream of eliminating beliefs in things that are beyond sensory experience. However, one of Quine's contemporaries, Rudolph Carnap, realized that the question about whether to adopt any linguistic framework is not answerable within the framework. It is a question that is external to the language. To decide whether to adopt a particular way of speaking about the world, one must specify how the way of speaking will be used. The justification for a particular linguistic framework ultimately depends on the framework's ability to achieve practical goals. Pragmatic factors, such as simplicity, efficiency, precision, and utility are among the decisive reasons for adopting a particular language (Carnap 1950).

Quine agreed with Carnap that the answers to metaphysical questions ultimately depend on linguistic choices that we make, which should be based on our practical goals. Quine did not use Carnap's linguistic framework terminology, however. Quine preferred to talk about theories that address our relationship to the world. According to Quine, idealism and realism are both philosophical theories that address our relationship to the world, which can be evaluated in the same way that we would evaluate any scientific theory. A theory is committed to existence of objects, properties, events, and processes that are implied by its descriptive statements. For example, if a theory contains the sentence "The sun is made of 90% hydrogen," it is committed to the existence of the sun and hydrogen. If a theory contains the sentence

"The sun fuses hydrogen to produce helium and electromagnetic radiation," then it is committed to the existence of hydrogen, helium, fusion, and electromagnetic radiation. The realist proposes a theory that is committed to the existence of mind-independent objects (and processes, properties, or events), while the idealist proposes a theory that is not committed to the existence of these mind-independent things. Both of these theories are compatible with information that we receive from our senses; the observations that we make do not determine which theory is the correct one. According to Quine, when two competing theories, whether from science or philosophy, are empirically equivalent, we should accept the theory that best promotes our goals. Factors such as simplicity, generality, testability, utility, and the like should play a key role in our decision to choose any theory.

Quine's ideas about empirically equivalent theories are based on the writing of twentieth-century physicist Pierre Duhem, who held that theories in physical science are underdetermined by the data. Quine expanded on this idea and applied it to all theories, scientific or otherwise. According to Quine, empirical observations do not uniquely determine theory choices; indefinitely, many theories are logically compatible with the same set of observations. Thus, a single test or even a series of tests cannot disprove one theory or prove another one because one can always modify a theory or one's background assumptions in order to accommodate recalcitrant evidence. Even though theories are logically underdetermined by the data, scientists can still make rational choices between different theories based on their practical goals (what they want to do) and epistemological goals (what they want to know). One may appeal to epistemological norms, such as explanatory power, fruitfulness, parsimony, and precision to decide between competing theories (or hypotheses). Because Quine and Duhem both developed this idea, the underdetermination of theories is known as the Quine-Duhem thesis (Klee 1997).

If we apply the Quine-Duhem thesis to the realism-idealism debate, we should accept realism instead of idealism because it has proven to be a very useful theory in ordinary life (Quine 1961, 1977, 1986). Consider how realism helps us to explain and predict events from ordinary life. Suppose there are three people who have sensory experience of what they regard as a chair. It is much easier for these people to explain and predict events related to these sensory experiences if they use the word "chair" in their conversations and understand that word as referring to a mind-independent object. Realism also offers us a simple, general, and powerful explanation of science's practical and epistemological successes, which we discussed above. Quine also uses this line of argument to justify positing the existence of abstract objects, such as numbers, points, and lines. We are justified in believing in abstract objects because they play an important role in mathematical and scientific reasoning. Quine's solution to the realism-idealism debate also answers the Kantian worry that the world might be "out there," but we cannot know it because Quine held that we are not justified in positing the existence of a mind-independent world unless we already have some evidence in favor of the theory that entails this commitment. For Quine, answers to metaphysical questions depend on the theory of knowledge, which depends on an understanding of how we use language to talk about the world (Quine 1961, 1977, 1986).

To summarize, Quine's pragmatic approach to metaphysical questions can answer the objection to the main philosophical objection to the realist strategy, discussed earlier. The main objection to this strategy is that it cannot succeed because there is no way that we can have knowledge of a mind-independent world. Quine argues that the question of whether we can have knowledge of a mind-independent world depends on what type of linguistic framework we decide to adopt. We could choose an idealist framework, which implies that we cannot have knowledge of a mind-independent world, or a realist framework, which implies that we can have knowledge of a mind-independent world. We are justified in adopting the realist framework because this framework helps us to achieve our goals in science and in ordinary life. We should use this framework because it yields results.

3.6 How Science Achieves Objectivity

We are nearing the end of this chapter's discussion of scientific objectivity. Some readers, who are not philosophers, may have very little patience with the complex and abstract questions that have addressed in this chapter, while other readers, who are philosophers, may find parts of this analysis to be overly simplistic. However, the investment of time and thought will be worth it. Far too many discussions of money's ill effects on science make sweeping generalizations that do not stand up to a more careful and critical analysis. This book will attempt to avoid this problem. It is important to have a clear idea of what we mean by "objectivity" in science before we can understand how money may undermine scientific objectivity. This chapter has attempted develop a general framework for thinking about objectivity that we can apply to specific cases, such as conflicts of interest or authorship and publication. Before concluding this chapter, it will be illuminating to make several comments about how science achieves (or fails to achieve) objectivity.

First, three different elements are responsible for objectivity of science: (1) scientists, (2) results, and (3) epistemological and practical norms. Results (including data, theories, and hypotheses) are objective insofar as they are unbiased or accurately describe reality; norms are objective insofar as they promote objective results; and scientists are objective insofar as they adhere to objective norms. Thus, the objectivity of science flows from scientists to norms to results. All these different elements are also important for achieving the objectivity of science. Science will not achieve objective results if the norms of science do not promote objectivity, or if the norms promote objectivity but scientists do not adhere to them. If scientists rely on results that are not objective, such as a theory based on biased data or assumptions, then additional results built on those biased results will also be distorted.

Individual scientists may be the weakest link the chain of objectivity, but they are an essential link nonetheless. Scientists may be the weakest link because they are human beings and may succumb to the temptations of money, power, success, and prestige, as well as financial, institutional, social or political pressures (Shamoo and Resnik 2003). In response to this concern, some writers have deemphasized the objectivity of scientists and have emphasized, instead, the objectivity of scientific norms or social organization (Solomon 2001; Longino 1990). Longino (1990) states that "objectivity, then, is a characteristic of a community's practice of science rather than an individual's" (74). Longino makes an important point here: we should think about the scientific community and its norms in understanding the objectivity of science. However, we should not overlook the importance of the individual. Scientific communities are composed of individuals. Individuals make decisions to solicit funding, design experiments, recruit subjects, perform tests, analyze data, interpret data, and publish results. A scientific community could not achieve objective results if most of the individuals who are members of that community did not adhere, most of the time, to norms that promote objectivity.

It is unlikely that even the best methods or the best community organization could overcome rampant data fabrication, carelessness, secrecy, or egregious conflicts of interest. If the vast majority of scientists did not strive to be objective most of the time, then science's social organization would fall apart and science would lose its public support and trust. For comparison, consider what would happen if most of the judges working in the legal system did not attempt to be fair most of the time. No matter how well designed a legal system may be, it still depends on individuals adhering to norms that promote its goals. The same point holds for just about any social activity, be it science or sports. Individuals play a very important role in sustaining (or destroying) human social activities.

It is also important to realize that objectivity in science is not all or nothing; it comes in degrees (Longino 1990). Each of the different elements of scientific objectivity—results, norms, and scientists—may be objective to a greater or lesser degree. A particular scientific theory, T1, may be more objective than another scientific theory, T2, because T1 provides a more accurate description of a mind-independent world than T2, or because T1 is not as biased as T2. A particular scientist, S1, may be more objective than another scientist, S2, because S1 more closely follows objective norms than S2. Finally, a norm, N1, may be more objective than another norm, N2, because N1 is better at promoting objective results than N2. Since each of the elements of scientific objectivity has gradations, the concept, as a whole, has gradations.

3.7 Conclusion

Most discussions of objectivity in science ask, "Is science objective?" This chapter has taken a different approach to objectivity in science and has asked, "Ought science to be objective?" The chapter has answered this question in the affirmative. Science ought to be objective because democratic societies need objective beliefs and methods to help resolve controversial moral, political, economic, cultural, and social debates. To help with the resolution of these debates, scientists should attempt to give unbiased testimony in public forums and should try to develop theories, hypotheses, methods and concepts that are free from personal, cultural, social, moral, or political biases. The most effective way of developing unbiased theories (hypotheses, methods, and concepts) is to attempt to test these theories against a mind-independent world. Granted, scientists often succumb to various biases, and it is often difficult to achieve objectivity in science. Nevertheless, the quest for objectivity is not an impossible dream or a vain pursuit. To achieve objectivity, the scientific community needs to develop norms that promote objectivity and individual scientists need to adhere to these norms. The previous chapter gave an account of some of these norms. The next chapter will consider how money can affect objectivity as well as other scientific norms.

FOUR

MONEY AND THE NORMS OF SCIENCE

When money speaks, the truth is silent.

—Russian proverb

4.1 Introduction

The previous two chapters have discussed the norms of science. Chapter 2 gave an account of those norms and chapter 3 focused on science's most important norm, objectivity. This chapter will explain how money can undermine the scientific community's adherence to these norms. Money can affect research by affecting the judgments, decisions, and actions of individual scientists and research organizations, including private companies, universities, government agencies, or professional societies. Money can induce individuals and organizations to make judgments and decisions that violate research norms, such as objectivity, openness, honesty, carefulness.

The violations of scientific norms may be intentional or unintentional. In the Nancy Olivieri case, discussed in chapter 1, the pharmaceutical company Apotex attempted to suppress her research showing that its drug had some harmful effects. This action by the company temporarily prevented Olivieri from publishing her data and results, which undermined the norm of openness. The company's conduct also prevented the scientific community from having important evidence relevant to the assessment of the drug, which interfered with the norm of objectivity. Finally, the company's actions could have also had a negative impact on the health of people taking the drug, which would have undermined the norm of social responsibility. In this case, the company intentionally violated ethical norms in order to avoid losing money. The scientific community usually reserves its harshest condemnations for intentional violations of scientific norms. People who intentionally break the rules may be guilty of fraud, plagiarism, harassment, theft, exploitation, and so on. Some intentional violations of ethical norms also violate laws and regulations.

Some scientists may believe that as long as they do not intentionally violate ethical norms that they have nothing to worry about. Nothing could be further from the truth. A person may make mistaken judgments or decisions as a result of subconscious influences on his or her thought and behavior. These influences may cause a scientist to fail to carefully examine specific assumptions or notice problems or anomalies. Although most scientists profess to be critical and skeptical, they are susceptible to self-deception (Broad and Wade 1993). In the Pons and Fleischmann case, also discussed in chapter 1, the researchers probably did not intentionally deceive the scientific community about cold fusion, but they may have made important mistakes and oversights as a result of their self-deception. In this case, money may have played a role in causing the scientists to succumb to self-deception. Money may have also had a subconscious influence on researchers involved in the Jesse Gelsinger case, also discussed in chapter 1, since James Wilson's financial interests may have caused him to underestimate the risks inherent in the gene therapy experiment. As we shall see in chapter 5, conflicts of interest can operate at a subconscious level. A researcher with a conflict of interest (COI) may sincerely believe that he is not biased, but he may still be biased. Everyone—even people with the highest degree of integrity—can be influenced in subtle ways by money.

Empirical research has provided us with a better understanding of the relationship between financial interests and ethical problems in science. For example, the literature on the relationship between the source of funding and research results, discussed in chapter 1, proves that there is often a strong correlation between the financial interests of the research sponsors and the results that are obtained and published (Krimsky 2003; Lexchin et al. 2003). However, proof of a correlation is not proof of causation. While these studies are very provocative, they do not tell us precisely how money affects the conduct of research. To show how money affects research, one needs to investigate all of the various links in the chain from research funding to research results in a particular case. To date, very few empirical or theoretical studies have carried out this type of analysis. A few authors have suggested some of the ways that money can lead to research bias (see, for example, Krimsky 2003; Greenberg 2001; and Resnik 2000a). Others, such as Wible (1998) and Kealey (1997), have developed economic models of scientific behavior, but they have not examined the relationship between money and scientific norms in detail. This chapter will explore some of the different ways that money can affect the practice of science and the public perception of science. It will use the steps of scientific research discussed in chapter 2 as a basis for understanding the relationship between money and research.

4.2 Problem Selection

Knowledge begins with a sense of wonder, and scientific research begins with the selection of a problem to study. Money plays a very important role in this beginning stage, since scientific curiosity is limited by the economic fact that it usually takes a great deal of money to conduct research. Gone are the days when scientists could fund their own research to study problems of their own choosing. The vast majority of scientists depend on government or industry contracts or grants to support their research and study problems that are of interest to those who fund research. While social and political factors play a major role in setting the government's R&D agenda, economic/financial factors play a major role in setting private industry's R&D agenda (Dresser 2001; Resnik 2001c; Resnik 2004a). Although private companies invest in some basic research, most allocate their R&D funds with an eye toward commercialization and profitability. To see why this is often so, it will be illuminating to examine the pharmaceutical and biotechnology industries in more depth.

In the United States, the FDA has the authority to regulate foods, drugs, biologics, and biomedical devices. The FDA can approve or disapprove a new product, or remove an existing product from the market. To obtain approval from the FDA for a new drug, a company must submit an investigational new drug (IND) application to the FDA and present data on the safety and efficacy its product. Before conducting experiments on human subjects, a company must have enough information from animal experiments to determine whether it is safe to use the product in humans. There are four different types of clinical trials that companies may conduct: Phase I trials, which test for the safety of a new drug on a relatively small number of subjects; Phase II trials, which test for safety and efficacy on a relatively small number of subjects; Phase III trials, which expand testing to a larger population of studies; and Phase IV trials (or post-marketing studies), which gather data on the long-term effects of the product and explore different use of the drug. In gathering data, the company must adhere to FDA standards for clinical investigation, which address issues relating to subject safety and rights, research design, and data integrity. The FDA examines the data provided by the company from its clinical trials and balances the benefits and risks of the new product. The FDA will allow the company to market the product if it determines that the benefits of the product outweigh the risks. The company only needs to demonstrate that the product is safe and effective at treating a medical problem or condition; the company does not need to demonstrate that the product is better than products that are already used to treat the problem, if there are any on the market (Goozner 2004).

If the FDA approves the new drug, it will grant to company exclusive rights to market the drug for a limited period of time, usually five years. Since patents last twenty years, and the process of drug development, including laboratory science, manufacturing quality control, and clinical trials, takes ten to twelve years, a company will usually have about eight to ten years of market exclusivity for its new drug, unless the company takes steps to extend its patents, which we will discuss in chapter 6. Once a patent on a drug expires, other companies are free to manufacture and market the drug under different names, since the company will still retain the right to control its trademarked name. Although companies may try to prolong their market exclusivity by filing lawsuits to keep competitors from making generic copies of their drugs, they eventually lose their control over the market. Patenting is usually the key pillar in market exclusivity, although a company may gain exclusive rights to market an unpatented drug under the Orphan Drug Act (ODA), which allows the FDA to grant exclusive rights for seven years to companies for drugs used to treat rare diseases (Resnik 2003a; Goozner 2004).

Because the FDA does not require pharmaceutical companies to prove that their products are superior to existing products used to diagnose, treat, or prevent a medical problem of condition, companies often develop many medications to treat the same problem. One company may develop a pioneering drug to treat a condition, but then other companies will enter the market with drugs very similar to the pioneering drug, or "me too" drugs. A good illustration of this phenomenon is a class of drugs used to treat depression and other mental health problems, known as the selective serotonin reuptake inhibitors (SSRIs). The market now includes many drugs with different names but similar chemical structures, such as Prozac, Zoloft, Paxil, and so on (Goozner 2004).

It costs a great deal of money to develop new drugs, biologics, or medical devices. According to some estimates, it takes, on average, $500 to $800 million to develop a new drug and bring it to the market (Goldhammer 2001). Angell (2004) and Goozner (2004) argue that these estimates of the price of drug development are highly inflated, and they place the average price of developing a new at far less the $500 million. If it only costs, on average, $300 million to develop a new drug, this is still a great deal of money. Even if a company successfully develops a new drug and brings it to the market, the drug may not be profitable: only about 33% of new drugs are profitable. Moreover, a company may have to withdraw a successful drug from the market due to safety or liability concerns (Goldhammer 2001). Thus, drug development is a highly risky business.

Pharmaceutical (and biotechnology) companies make R&D investment decisions based on a variety of factors that affect the profitability of a new product, such as the degree of intellectual property protection for a new product, the size and strength of the market for the new product, potential legal liability risks associated with the new product, market lead time and difficulty involved in manufacturing and marketing the new product (Goldhammer 2001; Angell 2004). Since they tend to make decisions based on these economic considerations, companies are more likely to fund R&D related to the treatment of diseases that affect people in the developed world (as opposed to the developing world), R&D on common diseases (as opposed to uncommon ones), or R&D on diseases that affect wealthy or powerful people (as opposed to impoverished or vulnerable ones). These economic decisions have an impact on the health of children, minorities, and people in developing nations. According to some estimates, 90% of the world's biomedical R&D funds are spent to treat 10% of the world's disease burden (Benatar 2000). Many medications that are prescribed to children have not been tested on pediatric populations. Although legal and regulatory difficulties with conducting research on pediatric populations help to explain the lack of research on children, economic factors also play a role. In order to provide pharmaceutical companies with an economic incentive to develop products that can help children, the U.S. Congress has enacted laws that extend patent protection by six months for drugs that are tested on and labeled for pediatric uses (Tauer 2002). Six months of extra patent protection may not seem like a long time, but it could mean hundreds of millions of dollars for a pharmaceutical company. As noted earlier, the ODA provides companies with incentives to develop and manufacturer drugs pertaining to rare diseases.

Many different potential biases are built into the U.S. government's system for approving new drugs, biologics, and medical devices. First, the FDA usually only examines data from studies sponsored by the companies submitting applications for new products. It usually does not examine independent data generated by other research sponsors, such as federal agencies. Even though the FDA requires that companies adhere to strict standards for research design, data collection, data analysis and ethics, these rules cannot overcome the potential bias inherent in this system because companies may find ways of working around or breaking the rules. Second, since the FDA does not requires companies to prove that their products are superior to existing products, its approval of a new product may sometimes not even benefit medical practice or public health, and may cost consumers more money, because physicians may use or prescribe new, expensive, patented products, rather than older, cheaper ones, whose patents have expired. For example, for many patients with high blood pressure, an older, diuretic drug, Lasix, may be just as effective as new, patented blood pressure medications (Goozner 2004). Third, since companies treat all data submitted to the FDA as confidential, business information, the agency does not make company data available to the public without permission. This policy makes it possible for companies to refrain from publishing adverse data presented to the FDA and can lead to biases in the research record (Angell 2004; Goozner 2004). Fourth, a company can publish the same data favorable to its product in more than one paper, which can skew the published research record. A systematic review of the published research will therefore overestimate the degree of evidential support for the company's product (Krimsky 2003). Fifth, companies require researchers to sign agreements giving the company the right to audit data, analyze and interpret data, control data, and review publications. These contracts give companies a great deal of control over the design of research, the analysis and interpretation of data, the ownership of data, and the publication of results (Angell 2004). Sixth, companies can control research outcomes by controlling the flow of money. A company can sponsor research on a study that it determines will benefit its products or it can withdraw funding from projects that appear to be generating undesirable results. If preliminary results show that a product is not safe or effective, a company may stop a clinical trial to protect subjects and to protect profits (Goozner 2004).

Research sponsored by pharmaceutical and biotechnology companies is fraught with potential biases. What should society do about these biases? First, the FDA should require, as a condition for approving a new product, that the company make all data it has gathered pertaining to that drug available to the public through a database, such as a clinical trial registry (Angell 2004; Shamoo and Resnik 2003). This would include data for products submitted to the FDA for approval as well as those that are withdrawn from the approval process, due to problems with safety or efficacy. The International Committee of Medical Journal Editors (ICMJE) now requires registration of clinical trials as a condition of publication (De Angelis et al. 2004; ICMJE 2005). Second, journals and editors should crackdown on duplicate publication—published the same data or results more than once—to prevent companies from skewing the research record (Huth 1986). Third, researchers and research institutions should pay careful attention to the contracts they sign with private companies to protect the integrity of the data and results and the researchers' publications rights. Fourth, government agencies should sponsor independent studies on new products to counterbalance potential industry biases. These studies could occur during or after the FDA approval process. If there are already effective products on the market, then the government-sponsored studies should compare the new products to existing products to determine whether the products are better than existing ones. Although the NIH currently sponsors some of these comparative clinical trials, it should conduct more of them. One promising idea is to form a new NIH center or institute for the improvement of clinical practice, which would design and conduct clinical trials aimed testing new clinical interventions (Goozner 2004). Fifth, researchers that conduct clinical trials for companies should assert more control over the planning, analysis, publication and interpretation of the research. Each of these different recommendations will be examined in more detail in this chapter and later on in the book.

It is worth noting that scientists did not always receive large sums of money from the government or private industry to conduct research. Ancient Greek philosopher/scientists, such as Plato, Aristotle, Euclid, Empedocles, Hippocrates, Archimedes, Hero, and Ptolemy, were teachers who also did research and scholarly work. They were supported by their students and other benefactors. In the twelfth century, European students organized the first universities and paid their professors' salaries. Professors who worked at these universities spent most of their time teaching and very little time on their own research or scholarship. They taught the classical liberal arts—grammar, rhetoric, logic, arithmetic, geometry, music, and astronomy—as well as the works from Greek, Roman and Arab philosophers, scientists, physicians, jurists, and historians, which had not been available in Europe during the Dark Ages (Haskins 1957).

Universities soon expanded their curricula and offered different scientific subjects, such as anatomy, medicine, biology, physics, geography, and chemistry. They began to receive more financial support from governments and from religious institutions. Many scientists supported their research with the income they earned as professors at universities. Andreas Vesalius (1514–1564), Galileo Galilei (1564–1642), William Harvey (1578–1657), and Isaac Newton (1642–1727) all held university professorships. Nicholas Copernicus (1473–1543) studied canon law, mathematics, and medicine at universities, but he was employed as a canon scholar and administrator in the Lutheran Church. Robert Boyle (1627–1691), who was independently wealthy, was able to fund his own chemical experiments. Robert Hooke (1635–1703) worked for Boyle and also for the Royal Society of London, the world's first scientific association. Newton earned more money working as the Master of the National Mint than he earned as a scientist (Meadows 1992).

Prior to the twentieth century, governments helped support research by supporting universities, but they rarely funded large research projects, with some notable exceptions. In 1572, King Frederick II of Germany gave Tycho Brahe (1546–1601) the island of Hven in the Sont near Copenhagen and paid for him to build an observatory there. In 1675, King Charles II built the Greenwich Observatory to help determine longitude, a very important measurement for navigation. Soon other countries with fleets of ships, such as France, built their own national observatories. Charles Darwin (1809–1882) held a university position, but he also did a great deal of his empirical research while serving as the ship's naturalist during the five-year voyage of the *HMS Beagle*. Gregor Mendel (1822–1884) was a monk who conducted his experiments on peas without outside funding other than what he received from the church for his subsistence. Albert Einstein (1879–1955) did some of his best work while working as an assistant at the Swiss patent office from 1901 to 1908. In World War I, governments invested money in military technology, such as aircraft and defenses against chemical weapons, but they cut back on science funding following the war. Government funding for research projects remained at a nominal level until World War II, which clearly demonstrated the importance of science and technology in warfare. During World War II, governments on both sides of the conflict funded research on airplanes, submarines, helicopters, radar, rockets, computers, cryptography, medicine, atomic energy, chemical explosives, and electronics. After World War II, the United States and the Soviet Union began investing heavily in scientific and technical research, especially defense-related research (Meadows 1992; Williams 1987).

One of the most important collaborations between academic scientists and industry took place in the late 1700s, when scientist-engineer James Watt (1736–1819) worked with entrepreneur Matthew Boulton to develop an efficient version of the steam engine. Watt developed and patented the steam engine, Boulton marketed it, and they both made money. Although the steam engine stands out as a striking example of science-industry collaboration, industry support for scientific and technical research did not have a significant impact on the funding of R&D until the nineteenth century, when German companies hired chemists to invent and develop synthetic dyes. Soon, other industries, such as the petroleum industry, the textile industry, and the food industry, began employing scientists, and the modern industrial laboratory was born (Williams 1987). By the 1930s, most of the R&D conducted in the United States was sponsored by private industry and took place in private labs. Bell Laboratories conducted more R&D than any university. Industry support for R&D has increased steadily in the last 100 years, and has been rising very rapidly since the 1980s, due to large investments in R&D by the biotechnology, pharmaceutical, computer, and electronics industries (Teitelman 1994). As noted in chapter 1, private industry currently sponsors about two-thirds of the R&D conducted in the United States and other industrialized nations.

4.3 Experimental Design

For the discussion of how money can affect experimental design, it will be useful to introduce a term coined by the nineteenth century mathematician and inventor Charles Babbage, who wrote about unethical practices he had observed in British science. Babbage (1830) distinguished between hoaxing, forging, trimming, and cooking data.

Hoaxing and forging involve making up data, and trimming involves omitting or deleting data that do not agree with one's hypothesis. Federal research misconduct regulations prohibit fabrication or falsification of data, which roughly correspond to hoaxing/forging and trimming (Office of Science and Technology Policy 2000; Resnik 2003b). Cooking the data occurs when one designs an experiment in order to obtain a result that one expects beforehand, for the purpose of affirming one's hypothesis. One could cook the data in order to obtain positive outcomes and to avoid negative ones. An experiment that cooks the data is not a genuine test of a hypothesis, but only a contrived test. A term not used by Babbage, fudging the data, involves the dishonest use of statistical methods to make the data appear better than they really are (Resnik 2000b).

Bias can arise in experimental design when researchers cook the data. There are many ways to cook the data, but perhaps the easiest way is to fail to measure a relevant outcome. In designing any experiment, one must decide which outcomes one will attempt to measure. If one decides not to measure a particular outcome, then one will not generate any data on that outcome. In experimental science, to be is to be measured. For example, suppose that a hypothetical company Pharmacorp is developing a protocol for studying the safety and efficacy of a new drug for lowering blood pressure. The company plans to measure many different variables, such as blood pressure, heart rate, respiration, temperature, blood cholesterol, blood acidity, blood iron levels, dizziness, nausea, kidney function, and so on. Also suppose that Pharmacorp does not plan to measure the drug's effect on potassium levels in the blood. If the drug can cause dangerously low levels of potassium, then the clinical trials conducted by Pharmacorp will not reveal this information. Even worse, suppose that Pharmacorp has conducted animal studies that give it some reason to believe that the drug could have adverse effects on potassium levels. If the company decides not to measure potassium levels in human subjects that take the drug, it would be guilty of cooking the data to avoid a bad outcome.

Another way to cook the data is to fail to collect enough data to detect a statistically significant effect, which may arise if one collects more data. Suppose that Pharmacorp decides to measure its drug's effects on potassium levels, but it does not collect enough data to demonstrate that the adverse effect of the drug is statistically significant. It designs an experiment so that it has enough data to demonstrate the good effects of the drug (with statistical significance) but not enough data to demonstrate bad effects with statistical significance. The experiment is under-powered with respect bad outcomes. If the bad effect occurs, but is not statistically significant, then the company will probably be able receive FDA approval for its drug, even though it might be required to tell consumers about the possibility of this bad effect when it markets the drug.[1]

If the FDA approves Pharmacorp's drug based on mistaken assumption about the risks of the drug, then Pharmacorp's cooked data will have a negative impact on the norms of science, as well as on society, since the drug could harm many patients unnecessarily. It is almost impossible to determine whether private companies ever cook their data, since their deliberations about research design are confidential. However, given the frequency with which new products are removed from the market after producing harmful effects in people, it seems likely that some cooking has occurred in many different industries.

Most of the industry-sponsored clinical trials today are designed by industry scientists instead of academic scientists. In a typical clinical trial, the company will select the research methods and the desired outcomes. The company will also write a protocol and an investigators brochure, which clinical scientists will follow in conducting the research. The typical clinical trial is company-driven rather than investigator-driven. The clinical researcher enrolls subjects and collects data, and may also be an author on a paper. For many years, clinical research was conducted almost exclusively at academic medical centers. Today, a large percentage of clinical researchers are physicians working in private practice. Many pharmaceutical and biotechnology companies hire contract researcher organizations (CROs) to manage clinical trials. The CROs recruit private-practice physicians to help implement the clinical trial and use private international review boards (IRBs) to oversee the research. Although many private-practice physicians have a genuine interest in contributing to the advancement of biomedical science, many of them are in it for the money (Angell 2004). Chapter 5 will explore these financial interests in more depth.

When a private company controls all aspects of research design, there is a great potential for bias. How can society combat this problem? The FDA's regulations pertaining to research design and data collection should help, in theory, to ensure the quality and objectivity of the experimental design and protocol. In addition, IRBs can exert some influence over experimental designs to protect human subjects from harm. For example, an IRB can require a research project to pass a scientific review before ethical review takes place. However, these organizations are far from perfect, and some flawed experimental designs can slip through. To avoid this problem, clinical researchers who are conducting industry-sponsored studies should assert greater control over experimental design. They should try to have some input into how studies are designed and implemented. They should encourage companies to make any changes that are necessary to ensure the quality and integrity of research or to protect human subjects. Researcher should use their bargaining power and should not allow themselves to be used by private companies.

Ideally, scientists should be the first line of defense against biased experimental designs. The norms of objectivity, honesty, and carefulness all imply that one should design experiments that do not cook the data. Indeed, deliberately and intentionally cooking the data could be regarded as a type of scientific fraud. But fraud may not be the biggest problem here. Much of the data cooking that occurs in science may be subconscious and unintentional. It is very easy to succumb to self-deception in science, especially when one has a financial interest in the outcome of research.

Perhaps the best way to deal with this problem is to ensure that a sufficient number of researchers without direct financial interest in the research results also conduct experiments pertaining to the same area of investigation. These independent researchers can serve as a check on the researchers who have financial interests that may be causing bias. Science conducted for the benefit of the public can play a key role in counterbalancing or buffering the potentially biasing effects of science conducted for the benefit private interests (Krimsky 2003).

4.4 Subject Recruitment

Before one can collect and record data in clinical research study, one must enroll participants in the study. Human subjects are the lifeblood of a clinical trial. If an investigator cannot recruit a sufficient number of subjects to yield enough data for a particular study, then he cannot complete the study or publish the results. Also, he may not be able to obtain the professional and financial rewards that come from completing the study, such as publication, intellectual property rights, additional funding, or an increase in stock values. Research sponsors also have a strong interest in recruiting subjects for clinical trials.

Indeed, some sponsors offer clinical researchers enrollment fees (or finder's fees) for recruiting subjects. Others compensate researchers for patient care costs, which include a substantial amount of money allocated to cover administrative costs. Clinicians can use these administrative costs to help support their practice and as an extra source of income. Drug companies usually pay clinical investigators several thousand dollars per patient to cover patient care and administrative costs (Shimm, Spece, and DiGregorio 1996; Angell 2004). When one factors in the salary support, honoraria, and consulting fees that sponsors offer to clinical investigators, these various forms of compensation add up to significant financial rewards for conducting industry-sponsored clinical research (Angell 2004).

There are at least two ways that these financial interests can affect subject recruitment. First, researchers may fail to provide patients with adequate informed consent if they have a strong financial interest in having the patient participate in research. Researchers may overemphasize the benefits of research and underemphasize the risks, and thereby contribute to the therapeutic misconception that is already common in clinical research. The therapeutic misconception occurs when a research subject mistakenly believes that a research study is a form a medical therapy designed to benefit him/her rather than a type of experiment designed to yield general knowledge (Lidz and Applebaum 2002). In the Gelsinger case, discussed in chapter 1, it is conceivable that Wilson downplayed the risks of the experiment and oversold its benefits. The experiment was a Phase I clinical trial, which means that subjects are not expected to benefit from the experiment. Gelsinger, though far from healthy, was able to follow a treatment regime that kept him medically stable. In retrospect, it might have made more sense to attempt the therapy on desperately ill infants with nothing to lose.

Second, researchers may bend or break the inclusion/exclusion criteria for clinical trials if they may gain financially from recruiting subjects. Most clinical trials have complex inclusion and exclusion criteria, which are needed to protect subjects from harm and to add a greater degree of control over the experiment. For example, a clinical trial on a new blood pressure medication might exclude people with very high blood pressure in order to protect them from the risks of an added elevation in blood pressure. Or the trial might exclude pregnant or lactating women to avoid risks to the unborn child or infant. The trial might also exclude people with cancer, kidney or liver disease, or a variety of other medical problems in order to reduce the number of uncontrolled variables. Violating exclusion criteria can harm patients and also invalidate research. In one highly publicized case, Roger Poisson, a Canadian surgeon conducting research as part of the National Surgical Adjuvant Breast and Bowel Project, falsified data on 117 patients from 1977 to 1990 in order to include them in the clinical trial (Resnik 1996). Although Poisson claimed that he violated the study's exclusion criteria so that his patients would have chance of qualifying for advanced treatment methods, one might question his motives.

How should society deal with these types of potential problems? First, disclosure of conflicts of interests in clinical trials would help improve the informed consent process. In their lawsuit against the University of Pennsylvania, Wilson, and other researchers, the Gelsinger family claimed Jesse Gelsinger would not have consented to the research study if had known specific details of the university's and the researcher's financial interests (Krimsky 2003). Second, closer monitoring and auditing of clinical trials would help to ensure stricter adherence to inclusion/exclusion criteria. Poisson violated the inclusion/exclusion rules for thirteen years. If the research sponsored or the IRB had reviewed his data and consent forms in a timely manner, they would have discovered this problem before it got worse.

4.5 Collecting and Recording Data

There are two different ways that scientists may violate ethical norms when they collect and record data. If a scientist unintentionally collects or records data improperly, as a result of sloppy or disorganized record keeping, this would be regarded as an error or mistake. If a scientist intentionally collects or records data improperly, as a result of dishonesty, this would be regarded as misconduct (Shamoo and Resnik 2003). As noted earlier, there are two types of misconduct related to collecting and recording data, fabrication and falsification. Plagiarism is another widely recognized type of misconduct, which we will discuss later. Poor record keeping is a serious problem in the ethics of research, but it is not nearly as serious a problem as fabrication or falsification of data. The difference between poor recording keeping and fabrication/falsification is like the difference between negligence and malfeasance.

Can financial interests have an impact on how scientists collect and record data? It is conceivable that a researcher might make an erroneous data entry, as a result of a subconscious desire to prove a particular theory or hypothesis. However, it is very difficult to determine whether this happens in science. People make mistakes all of the time. People also have wants, goals, or desires. From the fact that a person's mistake also promotes his or her financial interests, one cannot infer that the person's financial interests caused the mistake. Part of the meaning of a mistake is that it is not an intentional act. Explaining mistakes by appealing to subconscious desires, wants, or goals comes close to treating mistakes as if they are really intentional acts.

The connection between financial interests and fabrication or falsification is much easier to conceptualize and prove. In many of the confirmed cases of fabrication or falsification in science, significant financial interests were at stake (Shamoo and Resnik 2003). In many cases, if the offenders had gotten away with their misdeeds, they could have earned promotions or raises, or profited from investments or royalties. For example, from 1980 to 1983, Steven Breuning published twenty-four papers funded by a grant from the National Institute of Mental Health (NIMH). Breuning submitted fabricated data to the NIMH in his application to renew the grant for four years. A subsequent investigation by the NIMH found that he had fabricated and falsified data in reporting to the NIMH. Breuning was also convicted of criminal fraud, for which he was sentenced to sixty days in prison and ordered to pay $11,352 in restitution to the government. In 1981, a post-doctoral researcher at Harvard University, John Darsee, fabricated and falsified data relating to a drug therapy to protect against ischemic myocardium. If he had been successful, he could have profited from the therapy. In 1985, a Harvard cardiologist, Robert Slutsky, was being considered for promotion when 12 of the 137 publications on his vitae were found to be fraudulent (Shamoo and Resnik 2003). One could add even more examples to this list. It is likely that there are significant financial interests involved in most cases of scientific misconduct.

While it seems likely that financial interests play a significant role in causing researchers to commit misconduct in some cases, there are problems with establishing a general causal connection between financial interests and scientific misconduct. Although it is difficult to get a precise estimate of the rate of misconduct in science, the rate is probably very low. Nicholas Steneck (2000) used data from confirmed cases of misconduct to calculate the misconduct rate at 1 out of every 100,000 researchers per year. In one of the best surveys of misconduct, 6%–9% of faculty and students reported that they had observed misconduct or knew about someone who had committed misconduct (Swazey, Anderson, and Lewis 1993). Since most scientists probably have some financial interests at stake in research, and the misconduct rate is quite low, financial interests do not lead to misconduct in the vast majority of cases (99% or more). The overwhelming majority of scientists do not allow their financial interests to undermine the integrity of their data collection. Perhaps the best way to view the relationship between financial interests and misconduct is to say that financial interests do not cause misconduct, but they are a risk factor for misconduct (U.S. Congress 1990).

How should society respond to the possibility of misconduct or mistake when scientists have financial interests in research? There are several possible responses to this problem, including developing, publicizing, and enforcing conflict of interest policies (which will be discussed in more depth later in the book); developing, publicizing, and enforcing misconduct policies; developing, publicizing and enforcing policies on record keeping and data collection and management; and supporting and prioritizing high quality mentoring and training in science. We will discuss these responses later in the book.

4.6 Analyzing and Interpreting Data

Financial interests can affect how scientists analyze and interpret data. A researcher (or a company) with a financial interest in the outcome of the research may intentionally or unintentionally distort or "massage" the data to support that outcome. In the case involving Betty Dong, discussed in chapter 1, the company performed its own statistical analysis of Dong's results to try to disprove her conclusion that its hypothyroidism drug was no better than several competing drugs. There are many different ways that one could misuse statistics to arrive at erroneous or misleading conclusions. Some common misuses include (1) omitting data from a publication or presentation in order to develop a stronger relationship between two or more variables without discussing the reasons for the omission, (2) filing in missing data without discussing your methods for generating the missing data, (3) using an inappropriate method to analyze the data, and (4) using a misleading graph to represent the data (Resnik 2000b).

Consider the first type of misuse, which Babbage called "trimming." It is neither necessary nor desirable for a researcher to include all of the data in a particular experiment in the analysis of the data, since some of the data points may result errors or may be statistical outliers. For example, if a researcher studying a particular species of mold knows that a Petri dish has been contaminated with another species of mold, he or she can exclude these erroneous results from that dish from his or her dataset. A statistical outlier is a data point that is more than two standard deviations away from the mean. Statisticians have developed procedures for excluding outliers. It is acceptable to exclude outliers to clarify and strengthen relationships among variables, provided that exclusion does not change the overall results (Resnik 2000b). Suppose that a clinical researcher is attempting to show that there is linear relationship between the amount of a drug one takes (Y) and the clinical effect of the drug (X). In figure 4.1, the graph X, Y has two outliers: (a) and (b). A researcher may exclude these data points from the analysis in order to develop a stronger correlation between X and Y, provided that he or she discusses the reasons for excluding these outliers in any paper or presentation. The graph X1, Y1 also has several potential outliers: (a), (b), (c), and (d). A researcher should not exclude these data points because excluding them would change his overall results from a weak correlation between X1 and Y1 to a strong correlation. The ethical principles of honesty and objectivity require the researcher to not trim the data points in X1, Y1 or omit from the discussion. A researcher who has financial interests related to the outcome of the research could be tempted to omit the outlying data points in X1, Y1 from the analysis of the data.

Most definitions of research misconduct would treat intentional exclusion of outliers as misconduct, if the exclude affects the overall results (Resnik 2003b). The U.S. federal government defines research

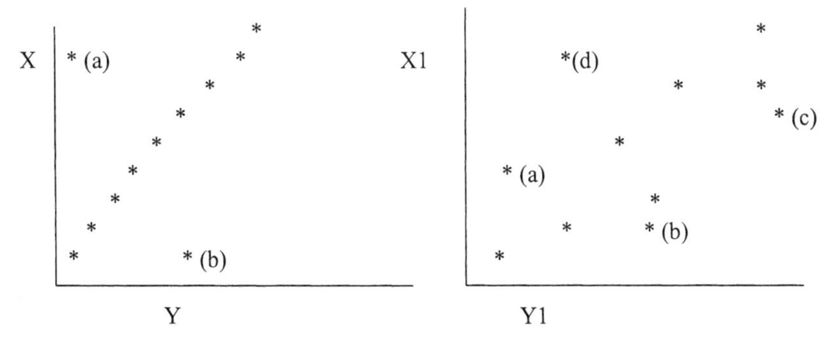

Figure 4.1. Excluding Outliers.

misconduct as "fabrication, falsification, or plagiarism in proposing, performing, or reviewing research, or in reporting research results." Falsification is defined as "manipulating research materials, equipment, or processes, or changing or omitting data or results such that the research is not accurately represented in the research record" (Office of Science and Technology Policy 2000, 76261). Under this definition, omitting or excluding data would be considered falsification if it results in an inaccurate representation of the research.

Financial interests could also affect a scientist's interpretation of the data. To interpret data, one must explain the scientific or practical significance of the data. Do the data disprove or prove one's hypothesis or a competing hypothesis? Do the data provide evidential support for any clinical applications or interventions? Since data interpretation requires considerable expertise and judgment, researchers will disagree about the interpretation of data, which leaves considerable room for aligning interpretations with financial interests (Als-Nielsen et al. 2003). For example, suppose a clinical scientist is conducting research on a new blood pressure medication and that he has financial interests related to the research. If the medication sells well, he will profit handsomely. The scientist could claim that the scientific significance of the medication is that it works through a novel mechanism of action in the body, and that the clinical significance of the drug is that it is more effective than currently used blood pressure medications. This scientist might intentionally or unintentionally exaggerate one or both of these claims about the drug's significance in order to promote his financial interests. He might, for example, claim that the drug is much more effective than competing drugs, when its effect on blood pressure is only 2% better than the effect of existing medications, and the new drug has some significant side effects. Alternatively, a company sponsoring the scientist's research might place some pressure on the scientist to interpret the data in a manner favorable to the company's product.

Two recent episodes illustrate how financial interests can influence data interpretation. On July 13, 2004, a panel of nine cardiology experts issued some guidelines for the use of cholesterol-lowering drugs. The drugs, known as statins, are sold under trade names, such as Lipitor and Zocor. The panel recommended lowering the thresholds for treatment with drugs to reduce the blood levels of low-density-lipoprotein (LDL) or "bad cholesterol." Their recommendation was based on a systematic review and interpretation of the data on the effects of cholesterol-lowering drugs. If the group's recommendations were implemented, seven million people in the United States would start taking these drugs, for a total cost of $26 billion per year. However, many cardiologists questioned the recommendations on the grounds that the benefits of taking cholesterol-lowering drugs must be weighed against the medical risks and the financial costs. The controversy reached a fever pitch when consumer groups exposed the panel's conflicts of interest. Eight members of the panel members were receiving money from the companies that manufacturer the drugs in question, but none of them had disclosed their financial ties to these companies (Kassirer 2004).

In 1990, the *New England Journal of Medicine* adopted a policy that prohibited authors of reviews or editorials from having any financial interests related to the products (or competing products) that they were discussing. The rationale for this policy was that reviews and editorials are highly susceptible to bias because they involve that analysis and interpretation of data from many different studies (Krimsky 2003). On June 13, 2002, the journal relaxed its policy after revealing that it had broken its own policy nineteen times from 1997 to 2000. The journal now prohibits authors of reviews or editorials from having significant financial interests (more than $10,000) related to the products they are discussing (Drazen and Curfman 2002).

How should society respond to problems with deceptive data analysis or interpretation when scientists have financial interests related to their research? Some possible responses include: (1) develop, publicize, and enforce conflict of interest policies (mentioned earlier); (2) develop, publicize and enforce policies on data analysis and interpretation; (3) sponsor independent (non-industry-funded) research. We will discuss these responses later on in the book.

4.7 Authorship

Before an article is published, the researchers involved in the study must decide who should be listed as an author and the order of listing. Financial interests can influence the allocation of authorship and credit in scientific publication. Tenure, promotion, and other personnel decisions made by universities are based, in part, on a faculty member's research productivity. Although many universities attempt to base their decision on the quality of publications as well as the quantity of publications, the dictum "publish or perish" still holds in the academy. Academic scientists have a strong financial interest in having their name appear on publications. The number of coauthors on scientific papers has been rising for several decades and continues to rise. For example, the average number of authors per article in biomedical journals has risen from slightly more than one in 1915 to more than six by 1985. One paper in quantum physics had 831 coauthors (Jones 2000). Although there are many different factors responsible for the steep rise in the rate of coauthorship, such as the complexity and scale of research and increased collaboration, financial and career pressures probably also contribute to this trend.

There are a variety of reasons why people are named as authors on scientific papers other than their contribution to the research. Honorary authorship occurs when someone is named as an author even though they have no made a substantial contribution to the manuscript (Flanagin et al. 1998). Honorary authorship may be granted in exchange for sharing data, materials, or reagents; as a personal favor; or as a gesture of gratitude or respect. In some cases a person may be named as a coauthor to give the paper added prestige or visibility. Some laboratory directors have insisted on being named as a coauthor on every paper coming out of their laboratory, and some colleagues have developed reciprocal agreements to name each other as coauthors on papers (LaFollette 1992). In one survey of 809 articles, 19% showed evidence of honorary authors (Flanagin et al. 1998). Honorary authorship is a serious violation of the ethical norms of science (Shamoo and Resnik 2003). Financial interests probably play a direct or indirect role in encouraging honorary authorship.

Ghost authorship is another violation of ethical norms in science related to financial interests in research. Ghost authorship occurs when an article fails to name someone who has made a substantial contribution to the article. Of 809 articles in one survey, 11% showed evidence of ghost authors (Flanagin et al. 1998). Pharmaceutical companies often regard publication as a form of marketing. If a paper supporting the company's drug is published in a scientific journal, then the company can cite that publication in magazine advertisements for consumers and brochures prepared for physicians. Many companies also disseminate copies of journal articles assessing their drugs to physicians. For marketing to physicians to be effective, it is important that the authors are affiliated with prestigious universities or medical centers. To ensure that respected names are on its publications, a company may hire a ghostwriter to write up the results of a study and pay researchers associated with universities or medical centers to serve as the listed authors. When this happens, the authors are authors in name only. In some cases, the named authors have not even read the manuscript (Flanagin et al. 1998). In one case reported in the media, Wyeth Pharmaceuticals signed a $180,000 contract with Excerpta Medica to acquire distinguished, academic scholars to serve as authors for its articles, which included editorials, reviews, and original research. Each article developed under this agreement paid $5,000 to a freelance writer to write the article and $1,500 to an academic scientist to be named as an author (Krimsky 2003).

In some cases, financial interests play a role in not giving credit where it is due. If a research project may generate financial rewards, scientists or private companies may seek to prevent collaborators from receiving proper credit on publications and patents. In some cases, senior researchers have stolen ideas from graduate students and patented the ensuing inventions. For example, in 1997 a jury awarded Carolyn Phinney $1.67 million in damages because it found that her supervisor at the University of Michigan had stolen her data (Shamoo and Resnik 2003). In another case, Jaony Chou sued her mentor at the University of Chicago for using her data on a variant herpes virus gene without her knowledge and developing a patented product from it. A federal appeals court in this case ruled that Chou had a right to assert her property interests in the patented invention (Marshall 2001d). In other cases, researchers have stolen ideas to support grants. In 1999, Antonia Demas filed a lawsuit against Cornell University nutrition professor David Levitsky for stealing her ideas on teaching nutrition and using them to obtain grants (Marshall 1999a). In still other cases, private companies have stolen ideas from researchers in order to prevent them from claiming intellectual property rights. For example, a federal court required American Cyanamid to pay Robert Allen and Paul Seligman $45 million in damages for stealing and patenting their formula for prenatal vitamins (Smith 1997).

In response to problems concerning authorship on publications, many scientific journals have adopted ethical standards for authorship. The standards adopted by the International Committee of Medical Journal Editors (ICMJE) are among the most influential ones:

> All persons designated as authors should qualify for authorship, and all those who qualify should be listed. Each author should have participated sufficiently in the work to take public responsibility for appropriate portions of the content. One or more authors should take responsibility for the integrity of the work as a whole, from inception

> to published article. Authorship credit should be based only on (1) substantial contributions to conception and design, or acquisition of data, or analysis and interpretation of data; (2) drafting the article or revising it critically for important intellectual content; and (3) final approval of the version to be published. Conditions 1, 2, and 3 must all be met. Acquisition of funding, the collection of data, or general supervision of the research group, by themselves, do not justify authorship. Authors should provide a description of what each contributed, and editors should publish that information. All others who contributed to the work who are not authors should be named in the Acknowledgments, and what they did should be described. (ICMJE 2005)

It is highly unlikely that all the papers published in scientific journals meet either these requirements or those adopted by other journals. Thus, many researchers are probably receiving credit for scientific publications even though they do not deserve credit, which can lead to a variety of ethical problems in science, including a lack of fairness and accountability (LaFollette 1992). Fairness in research requires that scientific rewards reflect scientific contributions: credit should be given where it is due (Pennock 1996). Accountability requires that researchers take responsibility for their contributions to a research project. A lack of accountability can pose serious problems for scientists when a publication contains serious errors or is suspected of presenting fabricated or falsified data. When this happens, it is very important to know who contributed to the publication and what they did. In numerous cases of scientific misconduct, coauthors have denied knowing anything about the disputed data or their collaborator's work. Those who take credit for research must also be prepared to share in the responsibility for research (Shamoo and Resnik 2003).

The U.S. Patent and Trademark Office (USPTO) (2003) defines an inventor as "one who contributes to the conception of an invention." Although this definition is better than nothing, it does not specify what type of activity might count as contributing to the conception of an invention. Is involvement in data collection sufficient for inventorship? What about experimental design or statistical analysis? Clearly, the scientific community needs to do more work in defining "inventorship" (Ducor 2000).

How can society deal with the ethical problems related to authorship and inventorship when financial interests are at stake? Many journals, professional societies, and universities have already adopted guidelines for authorship and inventorship. This step, while important, is not enough. Journals, professional societies and universities should clarify their guidelines, if they are unclear; educate researchers and research sponsors about their guidelines; and they should enforce their guidelines. Journals should require all coauthors to sign a statement indicating how they have contributed the work. Universities or professional societies should also consider conducting random audits to verify authorship status. (We will discuss authorship issues again in chapter 7.)

4.8 Publication and Data Sharing

When researchers complete their data-gathering activities for a research study and settle questions concerning data analysis and interpretation or authorship, the study is ready to publish. In most cases, researchers are eager to share their data and publish their results. However, financial interests can interfere with data sharing and the publication of research results. We have already discussed several cases in this book where scientists have faced pressures from companies to not publish their research or share data, methods, and tools. We have also mentioned some of the empirical studies that show that financial relationships with companies sometimes interfere with publication and data sharing in science. Blocking the publication of a scientifically valid research study interferes with openness in science and may also have negative consequences for society. For example, Apotex's decision to block publication of Nancy Olivieri's paper on some side effects of deferiprone may have harmed patients who were taking the drug. If DeNoble and Mele had been allowed to publish their research on tobacco, it is likely that people would have acted more quickly to regulate public smoking and to bring litigation against the tobacco industry.

Although critics of the privatization of research tend to view corporations as the main culprits when it comes to interfering with publication and data sharing, it is important to realize that corporations are not the only entities with financial interests that can be harmed by publication. Since universities are now in the technology transfer and development business, a university also may place pressures on scientists not to publish or share data, if publication or data sharing could undermine the university's intellectual property claims or business ventures. Since many individual scientists also have intellectual property interests or formed their own companies, they may decide not to publish or share data in order to protect financial interests (Shamoo and Resnik 2003). Finally, some scientists may refuse to share data so that they can protect their own professional and career interests. For example, if a researcher develops a large database and publishes one article from it, he or she may be reluctant to share that entire database with other researchers until he or she has finished publishing several additional articles from it.

Some writers, such as Krimsky (2003), Brown (2000), and Press and Washburn (2000), assume or imply that the pressure not to share data or results with the scientific community has arisen only in the last few decades, as a consequence of growing ties between academic science and private industry. Although the relationship between academic science and private industry poses a unique challenge for openness in science, the tensions that have arisen in the last few decades are not entirely new. For hundreds of years, scientists have struggled with the conflict between secrecy and openness. Most people know that Leonardo da Vinci wrote in mirror writing because he was worried about other people stealing his ideas. But what most people do not know is that Da Vinci was not alone in his desire for secrecy: for centuries, scientists have protected secrets because they feared that other people would steal their ideas. The development of scientific journals in the 1700s and 1800s played an important role in overcoming the desire for secrecy in research by providing scientists with receiving credit in public forums for their accomplishments (Shamoo and Resnik 2003).

Before countries adopted patent laws, many inventors kept their ideas and discoveries a secret in order to prevent other people from using, making, or commercializing their inventions. As we shall see later, the patent system involves a bargain between inventors and the government: the government grants the inventor a limited monopoly on his or her invention in return for pubic disclosure of information needed to make and use the invention. Even though industrialized nations have patent laws, scientist/inventors, companies, and universities can still protect their ideas through secrecy (A. Miller and M. Davis 2000). One reason that it is important to have clear, stable, and fair patent laws is to encourage people to pursue public disclosure instead of secrecy.

In addition to intellectual property interests, scientists have also kept secrets in order to establish priority. Priority of discovery is a very important concern for scientists even when patent rights are not at stake. In the early 1950s, three groups of investigators—James Watson and Francis Crick at Cambridge University, Maurice Wilkins and

Rosalind Franklin at Kings College (London), and Linus Pauling and his colleagues at the California Institute of Technology (Pasedena)—were locked in a race to discover the structure of deoxyribonucleic acid (DNA) (Mayr 1982). As we all know, Watson and Cricke won the race as well as the Nobel Prize. During this race, the different research groups did not share data freely, since they were each trying to be the first to discover DNA's structure. No patents were at stake in this contest, but secrecy still clashed with openness. For a more recent example, consider the bitter dispute that arose between Robert Gallo and Luc Montagnier over the discovery of the human immunodeficiency virus (HIV). Montagnier, who had collaborated with Gallo on HIV research, accused Gallo of stealing a virus sample in order to be able to claim credit for being the first person to isolate the HIV strain (J. Cohen 1994).

Scientists have also kept secrets to protect their research from premature publication. A researcher may not decide to share any data or ideas at a preliminary stage of the research in order to avoid professional embarrassment, to develop a strong argument before going public, or to avoid propagating erroneous data or results. These are all good reasons for sharing one's work only with a few trusted colleagues, but not with a public audience, until it is ready for public presentation. For example, Darwin waited over twenty years before publishing the theory of natural selection that he conceived of following the voyage of the *HMS Beagle*. He shared his ideas only with John Hooker (Mayr 1982). Darwin took a long time before going public with his theory because he knew that it would be extremely controversial and subversive. In the cold fusion case discussed in chapter 1, Pons and Fleischmann's press conference was a form of premature publication. The scientific community would have been better served if Pons and Fleischmann had spent more time reviewing their data, methods, and results before seeking publication.

Finally, scientists also have kept secrets in order to protect information that should be treated as confidential, such as human research data that identifies individual subjects, ideas and data disclosed during peer review, classified military information, information concerning personnel decisions, and so on. Sometimes scientists who submit their research for peer review may not disclose every detail to the reviewers. Scientists might describe their experimental protocol in their publication but leave off an important step or procedure, so that they can protect their work from being stolen (Shamoo and Resnik 2003).

Some writers have suggested that financial pressures to not share data or results may be hampering the progress of science (Heller and Eisenberg 1998). It is easy to understand, in theory, how secrecy might slow progress. If a researcher cannot obtain some data, results, reagents, tools, or methods that she needs to conduct a research project, then her research project will be stalled or stopped. Since people with secrets may want to limit the number of people who know their secrets, secrecy could also hinder collaboration in science. If collaboration is hindered, then the progress of science will be harmed, since collaboration is essential to progress. There is some evidence that there are problems with secrecy and data withholding in science (Blumenthal et al. 1997; Campbell et al. 2002).

However, the facts do not support this view of science as handicapped by a lack of sharing and publication; the facts support the opposite view. Science continues to make rapid and amazing progress despite financial (and other) pressures to maintain secrecy. One measure of progress—scientific publication—continues to increase at a near exponential rate. For the last two centuries, the number of scientific papers published each year has doubled every ten to fifteen years (Odlyzko 1995). For example, the American Chemical Society (ACS) published 4,500 pages in three journals in 1935. By 1995, the ACS published twenty-four journals and four magazines for a total of 200,000 pages of research and supporting information (American Chemical Society 1995). This increase represents a doubling of publication every twelve years. From 1994 to 1999, publications in the genetic sciences increased by 38%, even though the number of patent applications received by the PTO rose from less than 500 per year to 2700 per year (Resnik 2001b). A 38% increase in the rate of publications in five years represents a doubling in about thirteen years.

These reflections on the practice of science help us to have a clear, sober, and balanced understanding of the relationship between secrecy and openness in science. While financial interests can have a negative impact on publication and data sharing, they are not likely to completely destroy the ethics of openness or undermine the progress of science. Indeed, the evidence indicates that scientists continue to value openness, and science continues to make rapid progress, despite financial (and other) pressures to maintain secrecy in research. One reason that secrecy does not undermine progress is that it helps to protect intellectual property, which encourages innovation, investment, and entrepreneurship (Resnik 2003c). The best way for society to promote progress in science is to develop policies and rules for publication and data sharing in science and to educate scientists about these issues. We will discuss some of these policies later in the book.

By far, the biggest problem with secrecy is science is that secrecy can be used to distort the published research record and undermine the objectivity of science. We have already touched on this problem several times when discussing the Dong case, the Oliveri case, and instances where companies have attempted to suppress data that is unfavorable to their products. As mentioned earlier, requiring companies to register clinical trials in a public database can be an effective way of dealing with this problem.

Besides suppression of publication, there are other ethical problems in publication that may be due, in part, to financial interests. One of these problems is duplicate publication; some scientists have published virtually the same article in different journals without telling the editors (LaFollette 1992; Huth 1986). Duplicate publication is contrary to the norms of science because it is dishonest, wasteful, and can distort the research record (Shamoo and Resnik 2003). One reason that a person might be tempted to engage in this practice is that it is an easy way of increasing the quantity of publications that appear on one's curriculum vitae (CV). As noted earlier, the financial implications of this practice are straightforward, since personnel decisions in academia are often based on the amount of publications on one's CV. Another reason that a person (or company) might engage in duplicate publication is to attempt to skew the research record (Angell 2004). For example, suppose a company sponsors three clinical trials that produce positive results. The company could publish three papers from each trial, one original and two duplicates, for a total of nine papers. This would skew the research record and could make an outside observer believe that nine published papers supposed its product.

The other problem worth mentioning is publishing according to the least publishable unit (LPU), also known as "salami science." Publishing according to the LPU occurs when one divides a research study into a number of different publications (LaFollette 1992; Huth 1986). For example, a researcher adhering to this practice might derive nine publications from a research project, while a researcher not adhering to this practice might derive three publications. Since it is often appropriate to subdivide a research project to achieve simplicity and clarity in publication, some amount of judgment is required to determine how much a project should be subdivided. The person who practices salami science crosses the bounds of normal (acceptable) publication practice when it comes to subdividing projects. Publishing according to the LPU, like duplicate publication, is unethical because it is dishonest and wastes resources (Shamoo and Resnik 2003).

The financial motives for practicing salami science are similar to the reasons for duplicate publication: salami science can increase the number of publications on scientists' CVs with little extra effort, which can have an impact on personnel decisions about them. Additionally, a private company could publish according to the LPU in order to increase the number of publications favoring one of its products.

How should society respond to these last two ethical problems in publication that may result from financial interests? Once again, some scientific journals, professional societies, and universities have taken a first step by adopting polices that prohibit these practices. The next step is to support additional policy development as well as education for scientists. We will take up these matters again later in the book.

4.9 Peer Review and Replication

According to a traditional view, peer review and replication promote objectivity, truth, and the elimination of errors. Review by peers (a peer is someone who has knowledge or expertise in the discipline at issue) can improve the quality of published research and prevent scientists from publishing articles that contain errors or biases or that violate ethical norms. Peer review can also help to ensure that articles are original, significant, logically sound, and well written. Because scientists know that their work will be subjected to the peer review process, they are more likely to carefully scrutinize their own work before submitting it. Replication of results helps to validate published work. When an article is published, other scientists may challenge, interpret, or reproduce its results. If the results are irreproducible, there are probably some significant flaws with the research. Eventually, this system—peer review and replication—weeds out false theories and hypotheses and leaves behind only the true ones. Thus, science can be self-correcting (Popper 1959).

There are many different types of peer review in science. First, most scientific journals use scientific peers to review submitted manuscripts, namely, original articles, literature reviews, and so on. Second, most granting agencies use peers to review research proposals. Third, most academic departments use peers from outside the institution to evaluate candidates who are being considered for tenure or promotion. Fourth, some government agencies, such as the FDA and EPA, use peer review panels to make regulatory decisions.

Although peer review and replication play a very important role in promoting objectivity and other ethical norms in science, this traditional view is about the how science works is an oversimplification. The peer review system is far from perfect. First, reviewers often fail to catch simple mistakes, such as erroneous calculations or incorrect citations. Second, peer reviewers often are unable to detect biases, including biases related to experimental design and statistical analysis. Third, the peer review system is not good at detecting ethical violations, such as fabrication, falsification, or plagiarism. Fourth, peer reviewers are often themselves biased, unethical, or incompetent. Fifth, even when errors are detected, researchers often do not publish errata, and when they do, scientists often ignore the errata. Sixth, replication of results does not occur as frequently as one might suppose. Scientists rarely repeat experiments that are not their own work. Indeed, journals and granting agencies are usually more interested in new research rather than attempts to reproduce old research. Researchers usually attempt to replicate important results, but they may ignore other results. (For a review of these problems, see Shamoo and Resnik 2003; LaFollette 1992; Fletcher and Fletcher 1997; Chubin and Hackett 1990.)

Peer review has numerous flaws, but like democracy, it beats the alternatives. It is unlikely that eliminating the peer review system would enhance science's ability to adhere to normative standards, such as objectivity. The first scientific journals did not have peer review. The quality of published articles varied a great deal, including reports of experiments, rambling essays, and even poems. These journals initiated the practice of peer review to exert some degree of quality control over published articles (Shamoo and Resnik 2003). The need for quality control still exists. The advent of the Internet has made the need for quality control even more important in science, since the quality and reliability of information available over the Internet varies greatly.

There are no empirical studies concerning the financial interests of peer reviewers for journals or granting agencies. However, anecdotal evidence suggests that financial interests can cause ethical problems in peer review (Dalton 2001). The case concerning financial interests of the reviewers serving on FDA panels, discussed in chapter 1, illustrates vividly how financial interest could affect the objectivity of the peer review process. If a reviewer has financial interests related to an article or proposal that he or she is reviewing, then his or her interests might influence the recommendation. The reviewer may not even be aware of this influence on his recommendation: it may operate at a subconscious level. Since a significant percentage of scientists have research-related financial interests, it is likely that they also have financial interests when they serve as peer reviewers, since very often the same people who submit articles or proposals in a particular field also are asked to review articles or proposals.

Editors, review panel directors, and other people who make editorial or administrative decisions in peer review could also have financial interests. It is very important for those who control and manage the peer review system to be free from financial (or other) interests or pressures that could affect their judgment, decision making, or behavior. To illustrate this point, consider the following case. On January 15, 1999, the American Medical Association (AMA) dismissed George Lundberg as editor of the *Journal of the American Medical Association* (*JAMA*) for publishing an article on a study that surveyed college students about their sexual practices and beliefs (Kassirer 1999). The survey specifically asked whether the subjects regarded oral sex as "having sex" (Sanders and Reinisch 1999). The AMA claimed to have dismissed Lundberg for "interjecting *JAMA* into a major political debate that has nothing to do with science or medicine" (*JAMA* Editors 1999). One reason that the article was perceived as thrusting *JAMA* into a political debate is that President Clinton was under investigation for perjury relating to testimony he had given about an affair with White House intern Monica Lewinsky in which he claimed that they did not have sex. Many commentators criticized the AMA's decision as an infringement of editorial independence. They argued that the editor of *JAMA* should be free from pressure from the owner of the journal, the AMA (Kassirer 1999).

While it is not clear whether financial considerations influenced the AMA's decision to fire Lundberg, it is quite clear that political factors influenced the organization's decision. The AMA admitted this much. If political factors can threaten editorial independence, then financial interests (or some combination of both) can as well. One can easily imagine the AMA succumbing to pressure from a conservative lobbying group, a pharmaceutical company, or an insurance company. Problems like this can arise in scientific publishing because the owners of a journal may have financial or political interests that are at odds with the norms of science. Like many other scientific journals, *JAMA* is owned by a professional association. Other journals are owned by private publishing companies, academic publishing companies, private companies, or academic departments or schools. Some journals, such as *Science* and *Nature*, are not owned by an outside party.

Although editorial independence is important, the owners of scientific journals have a moral and legal right to define the aims and scope of their journals and to adopt editorial policies. For example, a journal dedicated to Jewish history has the right to refuse to publish an article alleging that the Holocaust was a hoax. However, even those journals with specific political or financial goals should still adopt policies that safeguard and define editorial independence (R. Davis and Mullner 2002).

How should society deal with these potential problems related to financial interests in peer review? The first step toward dealing with these types of problems is to define policies for dealing with financial interests in peer review. The NIH, the NSF, and most other funding agencies have rules pertaining to financial interests in grant peer review. Many journals have policies governing the financial interests of authors, but few have policies concerning the financial interests of reviewers and editors. The Committee on Publication Ethics (2005), which includes 337 scientific journals as members, has developed a code of ethics that addresses financial interests of authors, editors, and reviewers. Journals should consider adopting this code. Also, it is important to educate members of the scientific community about financial interests in publication and peer review and how to deal with them. We will take up these issues again in a later chapter.

4.10 Conclusion

It is no accident that there is a strong correlation between the source of funding and the results of research. Private and public sponsors of research pay scientists to answer questions that they would like to have answered. There is nothing inherently wrong with this arrangement. Practical, financial, political, humanitarian, and intellectual interests have always played a decisive role in problem selection in science (Resnik 1998a). Problems can arise, however, when financial interests intrude into experimental design, data analysis and interpretation, publication, peer review, and other aspects of science that should be protected from financial, political or other biases. When this happens, financial interests affect the process of scientific research; they can undermine objectivity, openness, honesty, and other research norms.

Although it is impossible to prevent money from having any impact on research, in order to prevent financial interests from undermining scientific norms, society should take some steps such as developing policies for journals, granting agencies, and research institutions; educating students and scientists about potential problems and issues; and monitoring research. Subsequent chapters in the book will explore some of these proposed solutions in more detail, and the last chapter will summarize the book's recommendations for scientists and society.

FIVE

CONFLICTS OF INTEREST

When Is Disclosure Not Enough?

> Money does not change men, it only unmasks them.
>
> —Mme. Riccoboni

Chapters 2 through 4 have laid the theoretical foundations for understanding how financial interests can impact scientific norms. The remainder of the book will explore concrete problems related to financial interests and practical solutions. This chapter will discuss conflict of interest (COI) in research; subsequent chapters will discuss intellectual property, publication, and government funding. Chapters 1 and 4 mentioned several examples of individual and institutional COIs but did not define or explain them. This chapter will begin by undertaking this expository work.

5.1 Defining Conflicts of Interest

In several of the cases described in chapter 1, such as the Gelsinger case, the Macknin case, and the Tseng case, the researchers had financial interests that may have affected their conduct in research. Consider a typical case of a COI in scientific research: a researcher owns a significant amount of stock in a company that is sponsoring his or her research. This type of situation is morally problematic because the researcher has some financial interests that could affect his or her conduct. As we saw in chapter 4, there are many different ways that these financial interests could affect the researcher's conduct. The researcher could introduce bias into the experimental design, fabricate or falsify data, misuse statistical methods in the data interpretation, suppress the publication of data or results, or violate ethical standards for recruiting subjects. All of these different improprieties would stem from problems with the researcher's judgment. Financial interests can lead to inappropriate conduct by causing judgment be biased, faulty, or unreliable (M. Davis 1982; Thompson 1993).[1]

The inappropriate actions taken by the researcher could result from deliberate choices and plans or subconscious motives and desires. Recent social science research on conflicts of interest suggests that self-serving biases in judgment can be intentional or unintentional (Dana and Loewenstein 2003). If a researcher intentionally violates widely accepted ethical, legal, or professional norms in research, his actions could be regarded as misconduct.[2] Although having a COI is not itself a form of misconduct, it may be a risk factor for misconduct (U.S. Congress 1990). If a researcher unintentionally violates these norms, his actions could be regarded as error or negligence. Making an error is not, by itself, inappropriate conduct, since errors and mistakes are a part of scientific discovery and validation. If a scientist makes an error and fails to use good judgment and reasoning to avoid the error, then this would be negligence. Negligence, or culpable error, is generally regarded as unethical in research (Shamoo and Resnik 2003). Figure 5.1 illustrates how personal, financial, or political interests can cause violations of ethical, legal, or professional norms.

Financial, personal, professional, or political interests that compromise (or could be perceived as compromising) judgment can also undermine trust. People will not trust scientists if they believe that the scientists' judgments have been compromised by financial or other interests. Even when financial, personal, professional, or political interests do not actually affect a scientist's judgment, they can still affect people's trust in that scientist because people might believe that the scientist's judgment could be comprised. Thus, COIs often boil down to matters of trust and trustworthiness (Resnik 1998c; De Angelis 2000). If people were perfect, then COIs would not be an ethical problem because people would always obey their ethical or legal duties and we could always trust them (Shamoo and Resnik 2003).[3]

Financial (or other) Interests → Compromised Judgment → Inappropriate Conduct (negligence or misconduct)

Figure 5.1. Financial or Other Interests and Inappropriate Conduct.

But people are not perfect; they may succumb to biases, temptations, prejudices, and they may make mistakes. Trustworthiness must be continually earned. With these thoughts in mind, we can now define a COI for a researcher:

> A *researcher* has a *conflict of interest* if and only if he or she has personal, financial, professional, or political interests that have a significant chance of compromising the judgment of the average scientist in the conduct of research.

A few comments about this proposed definition are in order. First, although many COI policies focus only on financial interests, the interests could also be personal, professional, or political. For example, a scientist who is reviewing a manuscript from one of his students for a journal would have a personal and professional interest at stake. A scientist with a strong commitment to the environmental movement might have a political interest at stake when he is evaluating a colleague for tenure and promotion. Second, the interests affect the scientist's judgment in the conduct of research. "Conduct of research" means, roughly, performing research activities according to accepted research norms. Thus, a scientist with a COI may exercise poor judgment vis-à-vis scientific norms, obligations, and duties. As a result of compromised judgment, the researcher may fail to adhere to epistemological, ethical, or legal standards for research (Resnik 1998c; Thompson 1993).

Third, the interests have a significant chance of compromising the scientist's judgment. The proposed definition does not say "compromise" or "appears to compromise." These semantic differences deserve our attention. Suppose we said that someone has a COI when his interests compromise his judgment. This definition would be too narrow. Under this definition, we would need to prove that a researcher's judgment has actually been affected by his interests to show that he has a COI. It would usually be very difficult to prove that a researcher's financial or other interests have actually affected his conduct in research (Krimsky 2003). It might turn out that we have enough evidence to prove that only a few researchers have COIs. Moreover, this narrow definition does not address the problems with distrust that are so important in understanding COIs. We would still have difficulty trusting a scientist even if we only believed that his interests could adversely affect his judgment.

Suppose we said that a person has a COI when her interests appear to compromise her judgment. The Association of American Medical

Colleges (AAMC) and the Association of American Universities (AAU) both use this type of definition. The AAMC defines COIs in science as "situations in which financial or other personal considerations may compromise, or have the appearance of compromising, an investigator's professional judgment in conducting or reporting research" (AAMC, 1990, 491). The AAU (2001) has adopted the AAMC definition. A chapter in a popular textbook on scientific integrity also defines COIs in terms of appearances: "A conflict of interest exists when an individual exploits, or appears to exploit, his or her personal gain or for the profit of a member of his or her immediate family or household" (Bradley 2000, 137). The problem with these definitions is that they are way too broad. There are many situations that may give the appearance of compromising a scientist's judgment, if by "appears" we mean "appears to someone" or even "appears to the average person." It is important, therefore, to distinguish between actual COIs and apparent COIs, and to not conflate these two ideas. COIs exist when an individual is in a situation that has a significant chance of compromising his or her judgment.

With these considerations in mind, we can also define an apparent conflict of interest for a researcher:

> A *researcher* has an *apparent conflict of interest* if and only if he or she has personal, financial, professional, or political interests that appear to the average outside observer to have a significant chance of compromising the judgment of the average scientist in the conduct of research.

This definition differentiates apparent COIs from actual COIs. One may have an apparent COI without having an actual COI, since to have an apparent COI one only needs to be in a situation that appears to compromise one's judgment. Actual COIs may also appear to compromise judgment, but they also are likely to compromise judgment. Apparent COIs may or may not be likely to compromise judgment. Although actual COIs are more serious than apparent COIs, scientists and research organizations should still develop policies for dealing with apparent COIs, since apparent COIs can still undermine trust. (We shall discuss some of those policies below.)

Before applying this analysis to institutions, it is important to resolve some other questions concerning the definition of a COI. First, what do we mean by "significant chance"? A 50% probability, a 20% probability, a 10% probability, or a probability greater than zero? In interpreting this phrase, we once again face the problem of making the definition too narrow or too broad: if "significant chance" means "at least 90% probability," then the definition will be too narrow; if it "a probability greater than zero," then the definition will be too broad. I think that most people would doubt the objectivity or integrity of a scientist's research if the chance that the scientist's financial or other interests will compromise his or her judgment is 5% or greater. I realize that this number is somewhat arbitrary, and some might argue for a higher or lower threshold for COIs, but it will have to do for our purposes here. One argument in favor of setting the threshold at 5% is that this corresponds to a popular significance level used in statistical reasoning.

Second, this proposed definition of a COI uses the *average scientist* as the reference class (or benchmark). Scientists may differ with respect to their ability to protect their judgment from personal, financial, or political interests. $100 may be enough money to affect some scientists, while $1 million may not be enough to affect others. For any particular scientist, it may be very difficult to know whether a financial (or other) interest is likely to compromise his or her judgment. However, it may be much easier to determine whether a financial interest is likely to compromise the judgment of the average scientist. Using the average scientist as the benchmark in this definition will prove to be very useful later in this chapter when we consider specific strategies for dealing with COIs.

Third, the proposed definition of an apparent COI uses the average outside observer as the reference class. The outside observer is someone who is not directly involved in the conflict but who would have knowledge of the conflict. This person could be a scientist or layperson. Since different people may have different reactions to conflict situations, it is desirable to use the average person, rather than any particular person, as a benchmark for this definition. For example, one person may believe that all conflict situations compromise judgment, whereas another person might believe that conflict situations never compromise judgment. Basing the definition of an apparent COI on either of these persons would be oversensitive or under-sensitive to apparent COIs.

Some of the cases described in chapter 1 also involved institutional COIs. For example, in the Gelsinger case, the University of Pennsylvania had financial interests related to gene therapy research. In the Olivieri case, the University of Toronto and Toronto General Hospital both had financial relationships with Apotex, the company that pressured Olivieri not to publish unfavorable results. In examining the FDA, we noted that this organization obtains a significant percentage of its budget from the very industry that it is charged with overseeing. Thus, it makes sense to ask whether research institutions, such as universities, government agencies, or professional societies, can have COIs. Institutions have all the necessary characteristics—duties, interests, and judgment—for having COIs.

First, institutions, like individuals, have a variety of ethical and legal duties to students, patients, faculty, and members of the greater community (Resnik and Shamoo 2002). Institutions have moral commitments to education, academic freedom, patient safety, knowledge advancement, and society (AAU 2001). They also must adhere to various laws and regulations pertaining to research. Second, institutions may have financial or political interests, such as ownership of stock or intellectual property, or contractual relationships with private companies or granting agencies. Third, although institutions are not conscious beings, they make collective decisions and have agents or employees who make decisions on their behalf. Institutions are incapable of judgment, perception, or reasoning, but they have processes for making decisions. A COI can compromise the integrity of those decision-making processes. For example, a university may decide to approve a particular contract, licensing agreement, or other financial arrangement with a private company. Or, an institution, acting through its IRB, may approve a human experiment. Fourth, administrators or other officials who serve institutions may also have financial interests that could compromise their judgment (AAU 2001; Moses and Martin 2001). For example, a vice president for research with a COI could attempt to pressure the IRB into approving a research study.

With these considerations in mind, we can define an institutional COI in research as follows:

> A *research institution* has a *conflict of interest* if and only if it has financial or political interests that have a significant chance of compromising the decision making of the average research institution in the conduct of research.

We can also define an apparent a COI for a research institution as follows:

> A *research institution* has an *apparent conflict of interest* if and only if it has financial or political interests that appear to the average outside observer to have a significant chance of compromising the decision making of the average institution in the conduct of research.

Readers will note that these two definitions for research institutions are almost identical to the definitions for researchers except that research institutions do not have any personal or professional interests. The interests at stake in institutional COIs are financial or political. A research institution may have financial goals, or objectives or even political ones, but a research institution does not have any personal relationships or professional goals.

Having defined COIs, it is important to define two related notions—conflict of commitment (COC) and conflict of duty (COD). A conflict of commitment is a situation in which a researcher has a secondary commitment that may compromise her ability to perform her obligations to her primary commitment. For example, if an academic researcher spends two days a week consulting for a private company, this would be conflict of commitment because (1) the researcher's primary commitment is to her academic institution; (2) the researcher also has a secondary commitment to the private company; and (3) this secondary commitment could prevent the researcher from performing her duties for the university. Although COIs and COCs both involve conflict, the difference between these two concepts is that a COC does not, by its very nature, compromise the researcher's judgment. The main ethical issue raised by a COC concerns the prudent management of a researcher's time and effort (Shamoo and Resnik 2003; Bradley 2000). However, a COC could become a COI if the researcher has some financial or professional interests related to her secondary commitment that are likely to compromise her judgment.

A COD is a situation in which a researcher has ethical or legal duties that conflict. For example, a researcher conducting a clinical trial might face a conflict of duties when deciding whether to stop a clinical trial that has shown very positive results or some serious adverse effects. On the one hand, the researcher has a duty to his patients to prevent harm to them and to provide them with the best available medical treatment. On the other hand, the researcher has a duty to the scientific community, future patients, and society to continue the clinical trial long enough to ensure that it will yield statistically significant results. If the researcher stops the trial too early, he may benefit his patients at the expenses of valid and significant research results (Schaffner 1986). Like COIs, CODs also involve conflict, but CODs do not, by their very nature, compromise the researcher's judgment. The main ethical problems with CODs are determining how best to balance conflicting obligations or duties (Shamoo and Resnik 2003). However, a COD could become a COI if a researcher has a financial or professional interest related to one of his conflicting duties that is likely to compromise his judgment.

5.2 Conflicts of Interest and the Norms of Science

How can COIs threaten or undermine the norms of science? Since COIs can compromise scientific judgment in the conduct of research, they can have a direct impact on a scientist's adherence to scientific norms. First, as we have noted several times, COIs can undermine adherence to the principle of objectivity because scientists and institutions with COIs may not take appropriate steps to prevent bias and may even introduce bias into research. Although most commentators focus on the relationship between COIs and bias, objectivity is not the only norm that COIs threaten.

Second, COIs can undermine honesty in science. By affecting a researcher's judgment, a financial (or other) interest could cause a researcher to fabricate or falsify data or commit plagiarism. A researcher might fail to disclose relevant information and assumptions about materials, methods, or statistical techniques. A researcher might stretch the truth on a grant application or in a scientific publication or presentation. Although there is no evidence that money causes dishonesty in research, many of the confirmed cases of scientific misconduct have involved significant financial interests (Shamoo and Resnik 2003).

Third, COIs can interfere with carefulness in science. A researcher with a COI may be less likely to scrutinize his own work critically. A researcher may want very much to believe that his hypothesis is true, if he may benefit financially from this state of affairs. The strong desire to believe that one's hypothesis is true may induce a researcher to underplay negative findings, overplay positive ones, and to succumb to self-deception (Kuhn 1970). The cold fusion episode, described in chapter 1, provides a useful illustration of this point. Strong financial interests in this case may have caused Pons and Fleischmann to be less than careful in their assessment of their research.

Fourth, COIs can threaten openness in science. Scientists with financial interests in research may be less likely to share data or publish data and results. Researchers may delay or avoid publication in order to protect their own intellectual property rights or the intellectual property rights of a private company. As mentioned in chapter 1, some researchers have delayed publication and refused reasonable requests to share data, tools, and methods. Researchers have also signed agreements with companies that give those companies the right to delay, review, and approve publications. Companies may pressure researchers to delay or avoid publication in order to protect intellectual property rights or to prevent the release of negative findings (Angell 2004). (Chapters 6 and 7 will discuss in more detail the ethical and policy dilemmas relating to intellectual property and publication.)

Fifth, COIs can interfere with the proper allocation of credit in science and respect for intellectual property. Scientists may fail to give credit to collaborators in order to promote their own financial or professional interests. Chapter 4 described several cases where researchers were accused of misappropriating ideas from students or colleagues. Money and career goals played a role in each of these cases. Scientists may also accept underserved credit in order to promote their financial or professional interests. Financial and professional interests clearly play a role in the problems of ghost authorship and honorary authorship described in chapter 4.

Sixth, COIs can undermine respect for human subjects in scientific research. As noted in chapter 4, a researcher with a financial interest in a clinical trial may be more likely to bend or break inclusion criteria, and may provide inadequate informed consent, in order to enroll a human subject in a trial, which could benefit the researcher financially and professionally. For example, a private company might give a researcher a finder's fee for enrolling subjects or provide the researcher with a generous reimbursement to cover patient care and administrative costs (Krimsky 2003).

It is also possible that a scientist with a financial interest in his or her research might also bend or break the rules for protecting the welfare of animals in research in order to accrue data. For example, suppose that a researcher is testing a new anti-cancer drug in animals. The researcher uses mice that are genetically engineered to grow cancerous tumors and injects them with the drug to see whether it affects tumor growth. In this experiment, the animals may grow very large tumors that cause pain or interfere with moving, sleeping, or eating. Animal welfare regulations and guidelines require researchers to euthanize animals to prevent unnecessary suffering (National Research Council 1996). However, euthanizing an animal could cause the researcher to lose valuable data if the animal does not live long enough to provide the data needed for the experiment. In order to ensure that he does not lose this data, a researcher with a financial interest in research could decide to keep the animal alive beyond the point at which the animal should be euthanized.

Seventh, COIs could also interfere with scientists' social responsibilities. In the Olivieri case described in chapter 1, Olivieri demonstrated great courage in standing up to Apotex. She believed that she had a social responsibility to report the dangerous side effects of the company's drug. She did this at considerable personal cost and risk to herself, and she almost lost her job, her reputation, and her career. In the Dong case, also described in chapter 1, Dong demonstrated considerable fortitude in attempting to publish results that Boots wanted to suppress. Dong faced costly litigation from Boots. How many other researchers have found themselves in a similar position and have decided not to oppose a company that wanted to suppress a negative result? Some scientists might add up the financial costs and benefits and decide that the most prudent course of action is to keep silent. Other researchers might eschew their social responsibilities in order to protect their financial interests.

Finally, COIs could undermine scientists' adherence to various norms relating to scientific epistemology, such as precision, testability, and empirical support, if researchers allow their financial interests to affect experimental design and data analysis and interpretation. These epistemological problems related to financial interests could also affect the objectivity of the research. In several of the cases described in chapter 1, financial interests may have caused scientists to violate epistemological norms. For example, Tseng violated empirical support when he minimized unfavorable results related to his research. Pons and Fleischmann violated the principles of testability and precision by failing to carefully control and describe the cold fusion experiments. Finally, the scientists on the FDA panels with COIs may have violated epistemological norms in their assessment of vaccines by basing their recommendations on insufficient evidence.

5.3 Strategies for Dealing with Individual Conflicts of Interest

Now that we have a better understanding of COIs and how they may affect the norms of science, we can discuss strategies for dealing with

COIs. We will begin with COIs for individuals and then consider COIs for research institutions.

There are three basic strategies for dealing with COIs that affect specific individuals: disclosure, prohibition, and conflict management (AAU 2001; Shamoo and Resnik, 2003). Disclosure involves informing relevant parties about one's financial or other interests. Parties would include anyone who might be directly affected by the COI, such as supervisors or employers, journals, granting agencies, audiences at scientific meetings, students, and research subjects. Prohibition involves preventing, prohibiting, or eliminating the conflict. For example, judges sometimes recuse themselves from legal cases when they have a personal or professional interest in the case. Politicians often place their investments in a blind trust when they take public office. Conflict management is more than mere disclosure but less than prohibition. Conflict management could involve the development of processes and procedures for overseeing and reviewing the COI to prevent or minimize any problems that it might create.

Disclosure is useful for at least two reasons. First, disclosure can help to counteract research bias, since it gives interested parties information they may need to evaluate the researcher's activities. For example, if a scientist discloses in a publication that he has received funding from a pharmaceutical company to conduct research and that he also does some paid consulting for the company, other researchers may consider the source of funding in evaluating the study. Readers with this information may carefully scrutinize the study for any potential bias.

Second, since disclosure involves honesty and openness (or transparency), it can build trust (AAU 2001). Disclosure gets the relevant information out in the open, for all to see. It also serves as a prophylactic measure because it avoids the potentially troubling situation of having one's financial or other interests disclosed by someone else. When this happens, people may assume that the person who failed to disclose a financial or other interest must have something to hide. Lack of disclosure can lead to suspicion and distrust. For example, in the Gelsinger case, discussed in chapter 1, Wilson and the University of Pennsylvania failed to adequately disclose their financial interests to the subject. Gelsinger's parents and others became suspicious when they found out about the financial interests at stake. In the Moore case, also discussed in chapter 1, Moore's physician also failed to disclose his financial interests, and Moore felt exploited when he found out about them.

Almost everyone agrees that disclosure is a very important way of dealing with COIs and apparent COIs. Granting agencies, such the as the NIH and the NSF, require disclosure if a researcher or his or her immediate family have a $10,000 equity interest in company that sells goods or services related to his or her research (Public Health Service 1995). Granting agencies also require reviewers to disclosure COIs. Universities have adopted COI policies to fulfill the requirements set by the NIH and NSF. Most of these university policies require some type of disclosure, although there is considerable variation in these policies concerning what must be disclosed (McCrary S V et al. 2000; Cho et al. 2000). Many scientific journals now require some type of disclosure of financial and personal interests by authors and reviewers (ICMJE 2005). The FDA also has COI disclosure policies (FDA 1999). Professional societies and associations recommend disclosure of COIs (AAMC 1990; AAU 2001; American Chemical Society 2000; American Physical Society 2002; American Society for Biochemistry and Molecular Biology 1998.).

Although there is a general consensus about the importance of disclosure in dealing with COIs, there are some key questions that need to be addressed concerning disclosure. To whom should disclosures be made? What (or how much) should be disclosed? At first glance, it would appear to be a good idea to disclose relevant information about COIs or apparent COIs whenever this information would be relevant to a party making a decision. But what determines relevancy? We could say that information about financial or other interests is relevant to a party if it would affect his or her decision, namely, if he or she would make a different decision as a result of the disclosure. The information would affect the party's decision if the party would reasonably believe that the researcher's interests have a significant chance of compromising his or her judgment. This definition covers most of the people who routinely receive COI disclosures, such as editors, granting agencies, employers, supervisors, institutional review boards (IRBs), and so on.[4] (Most granting agencies, journals, and universities require disclosure of significant financial interests. On the view defended here, an interest is significant if there is a 5% or more chance that it will compromise the judgment of the average researcher.)

One might also argue, from a moral viewpoint, that researchers have an ethical obligation to disclose financial (or other) interests to subjects, since this information could be relevant to the decision to participate in a clinical trial (Krimsky 2003; Morin et al. 2002).

Disclosure enhances autonomy. For example, the Gelsingers claimed that Jesse Gelsinger would have made a different decision if he had known about the specific financial interests related to the human gene therapy experiment that took his life. The Moore case, discussed in chapter 1, set the legal groundwork for disclosing financial interests to patients and research subjects during the informed consent process (Morin 1998). One might also argue, from legal perspective, that researchers have a legal obligation to disclose financial interests to subjects because a reasonable person would want to know whether a clinician or researcher has any financial interests that could affect his or her judgment. A reasonable person might refuse to accept a recommendation from a physician or clinical researcher if that person thinks that the recommendation is likely to be biased (Resnik 2004b).

While disclosure of COIs to research subjects makes a great deal of ethical and legal sense, some argue that investigators do not need to disclose their financial interests to subjects. First, some have argued that this information is irrelevant because it will have no impact on the subject's decision to enroll in a research study (T. Miller and Horowitz 2000). This argument is not very convincing because most patients would want to know this information (Kim et al. 2004). Jesse Gelsinger might not have enrolled in the gene therapy study if he had known about Wilson's and the University of Pennsylvania's financial interests. Subjects might not have been interested in this information in the past because paternalism reigned in medical practice and research, researchers did not have as many financial interests, and subjects were not as aware of the influence of money over clinical research. But times have changed. Paternalism no longer has a grip over medical practice or research, medical research is no longer conducted solely for advancement of medical science, and subjects have a greater interest in knowing how money might affect their medical treatment (Resnik 2004b).

Second, some have argued that disclosure of financial interests will make subjects suspicious and will undermine trust; It may be better to not bring up the topic. If subjects ask about financial interests, then disclosure would be appropriate, but if they don't ask, there's no need to disclose. According to Thompson: "Disclosure could even exacerbate some of the indirect consequences of conflicts, such as the effects on confidence in the profession or in the research enterprise . . . disclosure may merely increase levels of anxiety" (Thompson 1993, 575).

This, also, is not a good argument. Although disclosure may cause some subjects to become inappropriately suspicious, anxious, or distrustful, most will appreciate and understand the need for disclosure. People are accustomed to all sorts of disclosures today, including disclosures from realtors, stockbrokers, charities, accountants, businesses, and others. We live in a society where transparency is becoming increasingly important. Most research subjects will not be surprised or upset when financial disclosures are made in research. Indeed, investigators routinely inform subjects when a research study has a commercial sponsor. It is not too big a step from sharing this information to sharing other types of financial information. Researchers can avoid causing mistrust if they take enough time to explain to subjects the information that is being disclosed and why it is being disclosed. Good communication is the key to maintaining trust in the research enterprise (Resnik 2004b).

Third, some have argued that subjects won't understand financial information disclosed during the consent process (T. Miller and Horowitz 2000). Subjects will regard this information as similar to some of the legalese in the section on compensation for injury policy that one finds in a typical informed consent document. Since subjects won't understand the financial information, they will tend to ignore it. Hence, the information won't do them any good anyway.

It may be true that many subjects will not understand the financial information that would be disclosed; however, this is not a valid reason not to disclose. Many patients do not understand all of the risks of research or the various procedures that are part of the research protocol, yet investigators disclose this information. The proper response to this problem is to take enough time to explain the information to subjects and allow them to ask questions. It may also be necessary to provide subjects with some additional background information so they can understand why the information is being disclosed and how it might be relevant. Although it may take more time to do this, responsible investigators will be willing to take this step to ensure that informed consent is a meaningful dialogue between the investigator and the research subject, not just a signature on a piece of paper.

Currently, the FDA and NIH regulations require IRBs to request information about the financial interests from investigators and to assess this information when evaluating research proposals. Current regulations do not specifically require investigators to disclose financial interests to subjects. Even though IRBs collect information on financial interests, most do not require clinical investigators to disclose financial interest to research subjects during the informed consent process. Some have argued that IRBs should decide whether information should be disclosed to subjects on a case-by-case basis (Morin et al. 2002; AAU 2001). The strength of this approach is that it grants considerable latitude and discretion to IRBs. If an IRB determines that financial interests do not need to be disclosed to subjects, then it can refrain from requiring disclosure as a condition of approving a study. However, this strength is also a weakness because latitude and discretion could also lead to inconsistency and irresponsibility. The best approach is for IRBs to require that investigators disclose all significant financial interests to subjects, including sources of funding, and any financial interests greater than $10,000 (Resnik 2004b).

While disclosure is an essential strategy for dealing with COIs and apparent COIs, sometimes it may not be enough. There may be some situations in research that call for something more than disclosure, something such as prohibition or conflict management. Some COIs pose such a dire threat to science and society that most researchers agree that they should be prohibited. For example, granting agencies usually prohibit reviewers from reviewing proposals from collaborators, from researchers at the same institution, or from former students on the grounds that their personal or financial interests related to the review process could compromise their judgment. As noted in chapter 4, the *New England Journal of Medicine* prohibits authors of editorials and reviews from having a significant financial interest in the product of a company (or its competitor) being discussed. Some authors have argued that academic researchers should not serve as officers of a company or hold a significant equity interest in a company that is sponsoring their research (Krimsky 2003).

How should one decide whether a COI should be prohibited or merely managed or disclosed? According to the AAU, when it is necessary to protect the public interest or the interest of the research institution, an activity should be prohibited (AAU 2001). But what does it take to protect the public interest? This is not an easy question to answer, since responses to COIs involve trade-offs among different values. Although prohibiting a COI can benefit society, it can also have negative consequences. For example, suppose all journals prohibited researchers from having any financial interests in the articles that they submit for publication. This policy could promote objectivity and strengthen trust, but it could also deter researchers from publishing. Many articles that are now published would not be published if all journals adopted this policy, which would have a detrimental impact on scientific progress, and, by implication, the public interest. The wiser approach is to allow these articles to be published but to require researchers to disclose their financial interests. In making the decision to prohibit (or not prohibit) a COI, one should weigh the following factors:

1. *The significance of the COI.* As noted above, the significance of a COI is a function of its impact on the researcher's judgment. What is the probability that the COI would compromise the judgment of the average researcher? In general, the significance of the COI probably increases in direct proportion to the amount of money at stake or the closeness of a personal or professional relationship.
2. *The ability to manage (or control) the COI.* Sometimes it is possible to use various mechanisms to compensate for the biasing effects of the COI, but sometimes it is very difficult to compensate for these effects. If a researcher publishes an original article that reports data and results, then peer reviewers and readers are better able to detect and assess his or her bias than if the researcher publishes an editorial or review article. It may also be difficult to compensate for bias during grant reviews because reviewers may not be able to detect and assess bias in each other's reviews, or compensate for it.
3. *The consequences of prohibiting the COI.* Sometimes prohibiting a COI deprives science and society of important benefits. For example, suppose one prohibited scientists from serving on advisory boards or consulting with companies that sponsor their biomedical research. Although this prohibition might promote objectivity and trust, it would also deprive these companies of important expertise that they could use in making decision related to R&D. It would force researchers to choose between conducting research within an academic setting and serving a private company. All things considered, this type of prohibition could harm both science and business. Or suppose that one prohibited universities from receiving any research support from private industry. All academic researchers would need to be sponsored by the federal or state government or private philanthropies. Once again, this prohibition could have some good effects on science, but it could also have some bad effects. Forbidding universities from taking funds from industry would stifle research collaborations between academia and industry. These collaborations

can benefit both parties by allowing them to share information and expertise. (Bowie 1994)

With the above analysis in mind, I will briefly discuss whether there is a need to prohibit some specific types of COIs in research.

Peer review. In all types of peer review, including grant peer review and journal peer review, COIs should be prohibited. Researchers should not have a significant financial or other interest in a proposal or manuscript that they have been asked to review. Factors one and two, mentioned above, support this policy. In peer review, COIs can be very significant and may be difficult to manage. For example, a typical research article may only be reviewed by one or two editors and two or three reviewers. If one of those editors or reviewers has a financial interest that could be threatened by publication of the article or a grudge against the person submitting the manuscript, then his bias could have a significant affect on the outcome of peer review. Although prohibiting COIs in peer review may have some negative effects on science by disqualifying people with valuable expertise as reviewers, this is the price that one must pay for fair and unbiased peer review.

Research regulation/oversight. COIs should also be prohibited in research regulation and oversight activities such as review of human subjects research by IRBs; review of animal research by Institutional Animal Care and Use Committees (IACUCs); investigation of research misconduct by committees; COI review committees; FDA review of research sites; FDA and Office of Human Research Protection (OHRP) review of IRBs; FDA review of new drugs, biologics, and medical devices; and so on. Factors one and two also support this policy. When researchers have financial or other interests in research regulation or oversight, these interests may have a very significant impact on their judgment and may be difficult to manage. Although prohibiting COIs in research regulation and oversight may have some negative affects on the regulation of science by disqualifying people with valuable expertise, the need for fair and unbiased research regulation and oversight justifies prohibiting COIs.

Involvement with management of private companies. As noted several times in this book, many researchers today occupy management positions—such as president, vice president, chief executive officer, or research director—in private corporations that sponsor their research. Researchers who develop small start-up companies often serve as the president of the new company. One might argue that these types of relationships should be prohibited (Lo, Wolf, and Berkeley 2000; Gelijns and Their 2002). Researchers must choose between conducting research for a private company and helping to manage a private company. Let's analyze this problem in terms of the three factors mentioned above. The significance of this COI is very great because holding a management position in a company that sponsors one's research is likely to have a noticeable effect on one's judgment. The ability to control the COI is very low because there are usually few independent checks on managerial decisions made by companies. Finally, appointing scientists to management positions provides important benefits to companies, but the benefits are not worth the risks. Companies can take advantage of scientific expertise without sponsoring the research of a scientist who holds a management position in the company. For example, a company can consult with scientists or appoint them on advisory boards. Scientists can and should serve as presidents, vice presidents, and so on, but when they do, they should not conduct research for the company. They can enlist other people to do this job.

In some extraordinary cases it might be acceptable to allow a researcher with a major management position in a company to conduct research sponsored by the company. For example, suppose that the researcher is the person who is best qualified to conduct the research or occupy a key management position. In these situations, prohibiting the COI could undermine research or it could force the researcher to resign his or her management position. All of these outcomes would deny important benefits to the company, science, and society. In these rare cases, it would be better to not prohibit the COI but to attempt to manage the COI through a conflict management plan. For example, an independent committee could oversee and review the researcher's protocols, records, data, results, and publications.

Ownership of private companies. As noted several times in the book, researchers frequently own stock or equity in private companies that sponsor their R&D. James Wilson held a 30% equity interest in Genovo, for example. The problem with a researcher owning stock or equity directly related to his research is that the researcher might be tempted to violate ethical norms in order to increase the value of his stock/equity. For example, a researcher might publish biased data or manipulate patients into enrolling in clinical trials. The best way to avoid this type of problem is to prohibit researchers from owning any stock in companies that sponsor their research (Krimsky 2003; Gelijns and Their 2002). Most university policies only prohibit researchers from owning a significant amount of stock/equity in a company that sponsors their research, such as no more than $10,000 or a 10% interest. Obviously, the selection of a number that defines "significance" is a somewhat arbitrary choice. According to the analysis defended above, a "significant amount" of stock/equity would be the amount that would have a 5% or greater chance of compromising the judgment of the average researcher.

If one analyzes this problem by considering the three factors mentioned above, it would appear that ownership of a company that sponsors one's research can have a very significant affect on one's judgment. Moreover, it may be difficult to compensate for this biasing effect. Although the peer-review system can compensate for some of the biasing effects of research that the researcher decides to publish, it cannot compensate for research that he or she decides not to publish. For example, a researcher could decide to not publish negative results in order to avoid harming the value of her stock/equity. The peer-review system also may not be very effective at compensating for a researcher's biased interpretation of her data or results. Moreover, it is difficult to compensate for the biasing effect of stock ownership on the recruitment of subjects in clinical trials because research oversight bodies, such as IRBs, usually do not routinely observe or audit the informed-consent process. Finally, allowing researchers to own stock in companies that sponsor their research usually does not offer essential benefits to society.

The strongest argument for allowing researchers to own stock or equity is that this helps to finance a young company. Most companies when they first start out have difficulty raising capital and have a shortage of cash. Companies can raise capital by selling stock or equity, or by borrowing money (Hamilton 2000). A researcher who works for a new company could accept ownership of stock or equity in lieu of cash, which would allow the company to use its cash to pay other employees. Helping a company raise capital is a legitimate reason for allowing researchers (and universities) to own stock or equity in start-up companies. However, this argument does not support ownership of stock/equity directly related to one's research. A researcher who receives stock/equity as compensation for his services to a start-up company can still do this, but his services should not be services related to conducting research related to the company's products. The researcher could be compensated for contributing time and effort to developing the company's business plan, proposing R&D strategies, talking to potential investors, or making personnel decisions, for example. If the researcher owns a significant amount of stock/

equity in a company and intends to conduct research directly related to the company's products, then she should sell his stock/equity prior to conducting the research, or she should enlist someone else to do the research.

Sometimes allowing a scientist with a significant share of stock or equity in a company to conduct research sponsored by the company may be acceptable, if the scientists is the person best qualified to conduct the research. In these rare cases, it would be better to not prohibit the COI but to attempt to manage the COI through a conflict management plan, like those mentioned above.

Other financial relationships with private companies. Researchers may also work for companies as employees or paid consultants. They may also receive gifts or honoraria. All of these financial relationships could adversely affect a researcher's judgment and cause him or her to act in the company's interests. Although these types of relationships should be disclosed, they do not need to be prohibited. First, most of these other financial relationships can be compensated for in the peer review process or through the act of disclosure. Second, these other relationships can be managed through rules and policies that limit consulting activities and gifts or honoraria. Third, these relationships provide important benefits for private companies, which need access to scientific and technical expertise. In some cases, it may be appropriate to develop conflict management plans for researchers who have these other relationships with companies that sponsor their research, but in most cases disclosure will be sufficient.

Intellectual property. As noted earlier, many researchers have intellectual property interests related to their research, namely, patents or copyrights. These interests should be disclosed, and perhaps managed in some cases, but not prohibited. There are two reasons for adopting this policy. First, although intellectual property interests can have a significant affect on scientific judgment, it is often possible to control the effect of these interests. For example, the patent office functions like the peer review system because it requires inventors to disclose a great deal of information about their inventions and it will not grant a patent for an invention that does not work. Moreover, many inventions that are patented are also described in other peer-reviewed publications. Although there is no central office that reviews or grants copyrights, those who use copyrighted works provide a type of independent check on those works. For example, people will not use a copyrighted computer program that does not work well. Second, preventing researchers from owning intellectual property would have an impact on the advancement of science and technology, since it would remove an important incentive to conduct or sponsor research. (We will discuss intellectual property in more depth in chapter 6.)

Fees for enrolling patients in clinical trials. As noted earlier in this book, physicians have received various forms of compensation for enrolling subjects in clinical trials, including finder's fees such as $500 per subject enrolled, patient care costs such as $5000 per patient, and bonuses for meeting recruitment goals (Morin et al. 2002). I have already argued that these financial arrangements should be disclosed to IRBs and research subjects. Additionally, researchers should be prevented from receiving a significant amount of money for enrolling subjects in clinical trials. What is a significant amount of money? Again, this is difficult to define, but as a starting place we should say any amount of money that compensates the physician (or other health care professional) beyond his or her reasonable fees is significant. For example, if a company pays a physician $5000 per patient in patient-care costs, and the physician's actual costs are only $3000, including his or her time and staff time, then $5000 in patient-care costs would be $2000 in compensation beyond the physician's reasonable fees. This would be a significant amount of money. Likewise, a $500 finder's fee would be an unreasonable amount of money if it only takes the physician one hour to recruit and enroll a patient and he or she is normally paid at the rate of $200 per hour for treating patients.[5]

All of the three factors we have been discussing support a policy of prohibiting physicians from receiving a significant amount of money for enrolling subjects in clinical trials. They should be able to receive compensation for their time, but nothing more. First, the amount of money involved is likely to compromise the researcher's judgment. As we have noted earlier, many physicians regard clinical research as a source of income (Angell 2004). Second, it is difficult to manage this COI, since it is difficult to have an independent review of the informed consent process that takes place between the investigator and research subject or to audit the researcher's financial records. An IRB reviews consent forms, but does not usually monitor the process itself. An IRB asks for financial information, but usually does not audit financial records. Third, preventing researchers from receiving significant amounts of money for recruiting subjects would not have a detrimental impact on science or society. Investigators who are interested in conducting clinical research will still recruit patients without a large economic incentive. Only those investigators who are more interested in money than research would stop recruiting patients.

5.4 Strategies for Dealing with Institutional Conflicts of Interest

Concerns about COIs for research institutions are more recent than concerns about COIs for researchers. While there is a rich literature on the topic of COIs for researchers, much less has been written about COIs for research institutions. Federal agencies and professional organizations have had very little to say about the topic, and very few research institutions have developed or implemented policies or procedures for dealing with institutional conflicts of interest. This section will discuss some possible approaches for dealing with institutional COIs.

Disclosure is an important strategy for dealing with COIs for researchers and for research institutions. As noted earlier, disclosure can help to overcome bias, and it could build trust. Disclosure embodies the virtues of honesty, openness, and transparency. Accordingly, significant institutional conflicts of interest should be disclosed to the relevant parties, such as journals, granting agencies, and subjects in clinical trials. Various writers and organizations have proposed some strategies for managing institutional COIs (Resnik and Shamoo 2002; Moses and Martin 2001; AAU 2001), which I will discuss below.

Committees. Several organizations have recommended that universities develop COI committees to help manage individual and institutional COIs (AAU 2001). The responsibilities of these committees would be to solicit COI disclosures from individuals and institutional officials, review conflict management plans for individuals and institutions, develop COI policies, and sponsor educational activities related to COIs. Ideally, COI committees should include faculty from a variety of disciplines as well as non-faculty members (AAU 2001). A COI committee would function like an IRB, an IACUC, or other research oversight group. It could also share COI information with other oversight groups and facilitate the effective review of COIs by appropriate groups. Many research institutions have established COI committees (National Human Research Protections Advisory Committee 2001).

Although COI committees could play a very important role in dealing with individual and institutional COIs, one might argue that they will not be very effective at reviewing or managing institutional COIs. The problem is that these committees are composed of individuals who are employed by the research institution, and they are organized by people who are administrators within the research institution

(Resnik and Shamoo 2002). Even if a COI committee includes a few outside members, it will still be an institutional committee with very little authority within the institution, not an independent oversight body with substantial authority. If high-level institutional officials negotiate a deal with a private corporation that creates institutional COIs, it is doubtful that a COI committee would have the courage or authority to stop the deal. All a committee can do is quibble over the details of the COI management plan that comes with the deal. Genuine oversight of a COI can occur only when one has an independent group or person with the power and fortitude to manage or prohibit the COI.

There are several ways of responding to the problem. First, one could bolster the independence of the COI committee by adding outside members. If there are enough outside members, then people will be more likely to speak freely about potential problems with institutional COIs. Second, one could give the COI committee some actual authority by allowing it to approve or disapprove of licensing agreements, patents, contracts, and investments that create significant COI concerns for the institution. Third, one could also enlist the aid of an organization outside the researcher institution, such as the AAU, the AAMC, the NIH, or Public Responsibility in Medicine and Research (PRIM&R), to help review and manage institutional COIs. The outside body could establish a COI committee to review and manage COIs disclosed by its members. The committee could review these COIs and make recommendations to the university's COI committee.

If research institutions want to do more than pay lip service to the problem of institutional COIs, then they will take these three suggestions seriously. However, I doubt that this will happen without additional regulation or other legal repercussions. Universities and other research institutions rarely make radical changes in policy in the absence of outside legal, financial, or political pressures. Unfortunately, change may not happen until we have a few more high-profile tragedies like the Gelsinger case.

Firewalls. Another strategy for dealing with institutional COIs is to try to build firewalls within the institution to prevent financial or other interests from compromising decisions made by the institution (Moses and Martin 2001; AAU 2001). The institution should ensure that investment and licensing decisions do not influence the review of research. According to the AAU, "A key goal is to segregate decision making about the financial activities and the research activities, so that they are separately and independently managed" (AAU 2001, 12). Most research institutions already have firewalls in place. For example, a typical university will have not only an IRB to oversee human research but also an office of sponsored programs to negotiate contracts and grants with research sponsors as well as a technology transfer office to prosecute patents and negotiate licenses.

Firewalls are a good idea in theory. In practice, however, they are may be ineffective; it is often possible to get around them or go right through them because many of the key players within the research institution know each other and are familiar with the structure of the institution (Resnik and Shamoo 2002). In a typical university, the vice chancellor for research will know the chair of the IRB, the head of sponsored programs, and the director of the technology transfer office. These officials will probably also know many of the researchers who are pulling in large amounts of money through contracts, grants, and licensing, as well as leaders of private companies that have contracts with the university. In the real world, many of the key players within the research infrastructure at a university will know which research projects are likely to make money for the institution. It will often be difficult to prevent these institutional officials from influencing each other and from breaking down the firewalls.

One way of attempting to build a secure firewall would be to develop an independent, research institute associated with the university. The institution would be a legal and financially distinct organization, with its own bylaws and governing board. The research institution could hold intellectual property on behalf of the university as well as stock/equity. By placing investments and intellectual property in control of an outside organization, the university could separate its financial activities and research activities. This would be similar to placing investments in a trust. Many universities have developed independent research institutes to hold investments and properties, to sponsor R&D, and to support start-up companies.

Developing an independent research institute is also probably a useful strategy for dealing with institutional COIs, but it is no panacea. First, although a research institute would be legally and financially distinct from the university, it would still have a close relationship with the university. It would tend to promote the interests of the university, and the university would tend to promote the interests of the institute. For a relevant analogy, suppose that a researcher transferred all of his stock in a company or his intellectual property to a close family member, such as his wife or son. We can still say that the researcher would have a COI if he conducted research funded by the company because he would still have a financial interest (via his family member) that would be likely to compromise his judgment. In the same way, a university could still have a COI even if a research institute, closely connected to the university, holds its stock/equity or intellectual property related to research. Second, many of the key people in the research and financial infrastructure of the university would probably hold leadership positions in the research institute or have close relationships with people who hold leadership positions. They would deliberately or inadvertently go through or around the firewall created by the separate institution. Thus, concerns about institutional COIs could still arise when a university creates a research institute.

Policy development. Although most research institutions have policies for dealing with individual COIs, very few research institutions have developed policies for dealing with institutional COIs (AAU 2001). There are a couple of reasons for this discrepancy. First, discussion of the problem of institutional COIs among scientists, scholars, politicians, university leaders, and the public is a relatively recent phenomenon. People have been concerned with the issue of individual COIs since the 1980s, but were not very concerned about the issue of institutional COIs until 1999, when the media focused national attention on the Gelsinger case. Second, the federal government has issued regulations for individual COIs, but it has not issued policies on institutional COIs. Unfortunately, research institutions often avoid taking responsibility for research integrity concerns without a government mandate. The history of research ethics shows that research institutions have adopted policies on misconduct, human research, and animal research only after being required to do so by state or federal laws (Shamoo and Resnik 2003).

Clearly, policy development on institutional COIs can play a key role in dealing with institutional COIs. If policies are to be effective, they need the backing of the leadership at research institutions. Institutional leaders should make a commitment to following and implementing such policies, which would include educational initiatives as well as structural changes, such as the development (or strengthening) of COI committees and firewalls.

Prohibition. Some institutional COIs may pose such a dire threat to the institution's judgment, decision making, and integrity that they cannot be managed and should be prohibited (AAU 2001; Krimsky 2003). Recalling our earlier discussion of the factors to consider in deciding whether to prohibit a COI, one could argue that some types of institutional COIs should be prohibited because (1) they are significant, (2) they are difficult to manage, and (3) they do more harm than good. For example, one might argue that research institutions should not conduct clinical research when they have a significant financial interest at stake in that research, such as stock/equity or intellectual property rights. The University of Pennsylvania should not have conducted a gene therapy experiment in which it had a significant financial interest because (1) its COI was significant, (2) its COI was difficult to manage, and (3) its COI produced more harm than good. If a company has a significant amount of stock/equity or intellectual property rights related to a human research study, then it should enlist a different institution to conduct that study. The only exception to this policy should be if the research cannot be conducted at another institution because no other institution has qualified researchers or the appropriate facilities. Since the consequences of foregoing the research would probably be worse than the consequences of going ahead with it, the institution should attempt to manage the COI in such a situation.

One might also argue that a research institution should not conduct a human research study if a significant number of its employees occupy leadership positions in a company that is sponsoring the study on the grounds that this COI would be significant and difficult to manage, and the benefits of prohibiting the COI would outweigh the harms of not prohibiting it. In a situation such as this, the institution should allow another institution to conduct the study. This policy would also have an exception: if the research cannot be conducted at another institution, then the institution should attempt to manage the COI instead of prohibiting it.

Some writers have suggested that universities should be prohibited from holding stock/equity interests in research as well as intellectual property (Press and Washburn 2000; Angell 2000). They would like to reverse the trend toward technology transfer and commercialization that began in the 1980s with the passage of the Bayh-Dole Act. Although this suggestion would help to avoid institutional COIs in research, it could make the problem of individual COIs in research much worse. Under the current system, an inventor who works for a university can assign his or her patent to the university. If universities were not allowed to own patents, then the inventor would probably assign the patent to a private company, which might put additional pressure on the inventor to not publish or share data. Under the current system, if a researcher wants to start a new company, the university can provide the capital. If universities were not allowed to own stock/equity, then the inventor would have to obtain his or her stock equity from a private company, which might put pressure on the inventor to not publish results, introduce bias into the experiments, and so on. Even though university ownership of intellectual property or stock/equity in research raises serious concerns for the integrity of research, it may be preferable, in many cases, to ownership by a private company.

The only way to completely remove COIs that arise from these types of financial interests would be to prohibit individuals as well as institutions from having a significant amount of stock/equity in research or from owning intellectual property. But this option would have negative consequences for the progress of science, technology, and industry, since it would remove important incentives for innovation, technology transfer, and investment in R&D.

5.5 Conclusion

In this chapter, I have discussed COIs for individuals and for research institutions. I have considered three strategies for dealing with these conflicts: disclosure, management, and prohibition. I have argued for full disclosure of financial and other interests to relevant parties, which could include supervisors, employers, journals, professional audiences, IRBs, and even research subjects. I have argued that when disclosure is not enough, researchers and institutions should adopt strategies for managing COIs. I have also argued that it may be necessary to prohibit some types of COIs in some cases. For example, researchers and institutions should refrain from conducting human research when they have significant financial or other interests. Researchers should not review grant proposals or papers if they have a COI. They should also not participate in government advisory boards if they have a COI. Researchers should also not accept a significant amount of money to recruit subjects or conduct clinical trials. For the most part, I agree with the AAU's policy on COIs: "Disclose always; manage the conflict in most cases; prohibit the activity when necessary to protect the public interest or the interest of the university (AAU, 2001, ii). This chapter also discussed COIs related to intellectual property. The next chapter will explore intellectual property in more depth.

SIX

INTELLECTUAL PROPERTY

Balancing Public and Private Interests

He who owns the gold makes the rules.

—Old adage

We have discussed intellectual property in various places in this book. Chapter 1 mentioned several cases where intellectual property played an important role in questionable conduct in research, chapter 4 explored some ways that intellectual property can affect the norms of research, and chapter 5 considered how intellectual property can create conflicts of interests (COIs) in research. This chapter will explore intellectual property in greater depth and discuss its relationship to the progress of science and technology. It will argue that the best way to promote the advancement of scientific and technical knowledge is to seek an appropriate balance between public and private control of information. The chapter will begin with an introduction to the basic types of intellectual property.

6.1 What Is Intellectual Property?

Property can be understood as a collection (or bundle) of rights to control some object. For example, a person who owns a house has rights to possess, use, sell, lease, alter, or destroy the house. The house is the object; his property is his collection of rights over the house. Some properties give people rights to control over tangible objects, such as houses, lands, cars, cattle, and so on. Other properties grant people rights to control intangible objects, such as poems, songs, computer programs, stocks, and bonds. Intangible objects do not have any particular location in space or time; they are abstract ideas, not concrete things. The legal system recognizes two basic types of intangible property: financial instruments, such as stocks and bonds; and intellectual property, such as useful inventions and original works. All intellectual properties are naturally nonexclusive; two people can use the same item of intellectual property without diminishing each other's ability to use it. For example, two people can both read the same poem at the same time, but they cannot both use the same toothbrush at the same time. Because intellectual properties are naturally nonexclusive, the legal system allows people to obtain exclusive control over intellectual property. There are four traditionally recognized types of intellectual property rights (IPRs): patents, copyrights, trademarks, and trade secrets (A. Miller and M. Davis 2000).

6.2 Patents

A patent is a right granted by the government to an inventor to exclude other people from making, using, or commercializing his or her invention. An inventor can license other people to use, make, or commercialize his or her invention, and an inventor can assign his or her invention to another party, such as an employer. A license agreement is a legal contract between the owner of the patent and the person seeking to use, make, or commercialize the invention. Companies in technology-intensive industries, such as biotechnology and consumer electronics, often sign licensing agreements for each other's inventions, known as cross-license agreements (CLAs). Some of these agreements, known as reach-through license agreements (RTLAs) give the licensee the right to develop new products from the invention. In the United States, an inventor could also decide to keep his or her invention on the shelf, and allow no one to make, use, or commercialize it. The United States does not have compulsory licensing laws. European countries, on the other hand, have compulsory licensing laws. If the inventor does not make, use, or license the invention or license other to do so, then the government may step in and license the invention to another inventor. In the United States and many other countries, the term of a patent lasts twenty years from the time that the person files the patent application (A. Miller and M. Davis 2000).[1] Patents are not renewable; when the patent term ends, other people do not need the permission of the inventor to make, use, or commercialize the invention.

The knowledge disclosed on the patent application becomes available to the public when a patent is awarded. The U.S. government, for example, makes patent applications available on its website (www.uspto.gov). A patent, then, is a type of bargain between the inventor and the government: the inventor publicly discloses the invention in exchange for exclusive control of the invention. Public disclosure of scientific and technical information through patents benefits science and society. One million new inventions are patented each year. Patents are the world's largest body of technical information (Derwent Information 2001).

A patent application may include one or more inventors. Under U.S. law, an inventor may patent a machine, an article of manufacture, a composition of matter, a process, or an improvement on any one of these (U.S. Patent Act 1995). An inventor may also patent a new use of an invention (A. Miller and M. Davis 2000). For example, one could patent a light bulb, a process of manufacturing a light bulb, a new use for a light bulb (such as using a light bulb to cook hot dogs), and an improvement on a light bulb (such as a longer lasting bulb). In the application, the inventor must disclose enough information to allow someone trained in the relevant discipline or practical art to make and use the invention. The inventor need not provide a model or sample of his invention; a written description will suffice. The application must also provide some background information and make claims about proposed inventions. A claim describes the item that inventor wants to patent. For example, a claim might state, "An incandescent light bulb with a tungsten filament." During the application process, the patent office may challenge the inventor's application, and the inventor may provide additional evidence to support it. If the office does not award the patent, the author may submit a new or revised application.

In deciding whether to award a patent, the U.S. Patent and Trademark Office (USPTO) will determine whether the proposed invention satisfies the following necessary conditions:

Patentable subject matter. To be patented, an invention must qualify as a patentable subject matter. U.S. courts have ruled that laws of nature, natural phenomena, wild species, mathematical formulae, and abstract ideas cannot be patented. For many years, the U.S. legal system did not allow patents on life forms, except plant varieties. In *Diamond v. Chakrabarty* (1980), the U.S. Supreme Court ruled that Ananda Chakrabarty could patent a genetically engineered bacterium because the organism was the product of his ingenuity.

Chakrabarty had developed, using recombinant DNA techniques, a strain of bacteria that metabolizes crude oil. The bacteria could be used to help clean up oil spills. The court said that Chakrabarty's bacteria were patentable because he had produced a useful product with characteristics unlike any found in nature. It was his handiwork, not nature's.

In 1980, the USPTO awarded Stanley Cohen and Herbert Boyer a patent on a laboratory method for cloning recombinant DNA (S. Cohen and Boyer 1980). Chakrabarty's patent and the Cohen-Boyer patent played an important role in securing intellectual property protection for the emerging biotechnology industry. Since 1980, the U.S. government has awarded patents on DNA, RNA, proteins, hormones, cell lines, microorganisms, genetic tests, gene therapy techniques, recombinant DNA techniques, genetically modified plants, and mice. Other governments have followed the United States' example. Scientists and citizens have raised various moral and scientific objections to the patenting of biological materials (Resnik 2001b, 2003a, 2003g, 2004c).

Novelty. The invention must be new and original: it must not have been publicly disclosed through prior publications, public uses, sales or patent applications. To determine whether an invention is novel, the USPTO will examine the prior art to determine whether the invention has been disclosed in the prior art. Most would-be inventors maintain strict secrecy prior to submitting a patent application to avoid disclosures that could jeopardize their patents. If an inventor does not pursue his invention with due diligence, and a competitor learns about it, then the competitor could patent the invention (see below). Moreover, the inventor's own publications count as public disclosures. Because it is often difficult to know whether a publication will jeopardize a patent, most scientists who are interested in patenting do not publish or share data/results until they file a patent (A. Miller and M. Davis 2000).

Nonobviousness. The invention must not be obvious to a person trained in the relevant discipline or practical art. To ascertain whether an invention is obvious, the USPTO will examine the prior art and decide whether a practitioner of the art would regard the invention as an obvious implication or consequence of the prior art (A. Miller and M. Davis 2000).

Usefulness. The invention must have a definite, practical use, not a hypothetical or speculative use, a throwaway use, or a use as mere research tool. A patent is not a hunting license (*Brenner v. Manson*

1966). In the United States, the USPTO may deny a patent application if the invention only has an illegal use. In Europe, patent offices may reject an application if the patent has a use contrary to public morality.

Due diligence. In the United States, the first person to conceive of an invention will be awarded the patent, unless he or she does not pursue the invention with due diligence. An inventor has a one-year grace period between conceiving his invention and filing a patent. If he does not file the patent within that period, he could lose the patent to another inventor who reduces the invention to practice and files an application first. In Europe, the first person to file a valid patent application, not the first person to conceive of an invention, is awarded the patent. Therefore, in Europe there is a race to the patent office following invention.

Infringement. Once an inventor is awarded a patent, other people or organizations that make, use, or commercialize the invention without permission could face a patent infringement lawsuit. If an inventor suspects that someone is infringing her patent, she must file a lawsuit in federal court. The judge may then issue a temporary injunction barring the defendant from engaging in the alleged infringing activity. During this legal proceeding, the plaintiff may present evidence of infringement, and the defendant may counter this evidence with his own evidence.

In this case, a key issue before the court is whether the defendant used, made, or commercialized an invention that was substantially equivalent to the plaintiff's patented invention. Two inventions are substantially equivalent if they perform the same function in the same way to achieve the same result (A. Miller and M. Davis 2000). The courts have some leeway in interpreting the phrase "substantially equivalent." If they give it a loose reading, then the scope of the patent will be broad. If, on the other hand, they give it a very strict reading, then the scope of the patent will be narrow. If a patent has a broad scope, the inventor has much more control over the market than if the patent has a narrow scope. Courts must balance these sorts of issues when hearing patent infringement cases. Many courts are willing to give fairly broad scope to an invention that opens up a new field—a pioneering invention—in order to reward scientific and technical innovation.

Most inventors, companies, and universities want to avoid patent litigation, which can be expensive and time consuming. For example, the University of California spent over $20 million in a nine-year legal battle with Genentech over its patent on Protropin. The university had sought $4 billion in damages, but the parties settled their case for $200 million (Barinaga 1999). In order to avoid infringement, many companies in high-tech industries, such as electronics, biotechnology, and information technology, sign licensing agreements. Some authors have argued that problems related to the licensing could undermine discovery and innovation in biotechnology (Heller and Eisenberg 1998). These authors claim that patents on upstream technologies can inhibit the development of downstream technologies due to problems with negotiating licenses and high transaction costs in research, such as legal services and licensing fees.[2] In some cases, it might be possible for a patent holder to block the development of downstream inventions by refusing to grant licenses (Guenin 1996). Those who support licensing agreements, however, argue that companies have strong economic motives to negotiate licenses, that transaction costs are not a prohibitive cost of doing business for most companies and organizations, and that blocking patents is rare in biotechnology because patent holders have strong economic motives to license their inventions (Resnik 2004c).[3]

The research exemption. There is an important, though rarely used, exemption to patent infringement known as the research (or experimental use) exemption. Although the exemption is not a part of the patent statutes, the U.S. and European courts have recognized this exemption as a defense to an infringement lawsuit (Karp 1991). In the United States, the courts have construed the exemption very narrowly. The exemption only applies when the alleged infringing use of a patented invention is for research purposes, not for commercial purposes. For years, many universities have operated under the assumption that they can take advantage of the research exemption because they are interested in research, not practical applications. A recent case, *Madey v. Duke University* (2002), challenges this assumption. In this case, the inventor sued Duke University for using some of his laser technology without his permission. Duke argued that its use of the invention was covered by the research exemption, but the court rejected this argument on the grounds that the university's use was not a pure research use because the university was also interested in commercial applications. The court's decision in *Madey* reflects the realities of modern research, where the lines between commercial and non-commercial research have become blurred (Resnik 2003d). In order for academic researchers to benefit from the research exemption, it may be necessary to amend the patent laws to clarify and strengthen the exemption (Eisenberg 2003; Resnik 2004c).

6.3 Copyrights

A copyright is a right granted by the government that gives the author of an original work the right to exclude others from copying, performing, displaying, distributing, commercializing, or deriving works from an original work without permission. Copyrights last the lifetime of the author plus 70 years. If the original work is a work for hire, copyright protection lasts for 95 years from publication or 120 years after creation. Unlike patents, copyrights are renewable. If a copyright is not renewed, the work becomes part of the public domain. Authors may transfer their copyrights to other people or organizations or license others to use their original works (A. Miller and M. Davis 2000).

An original work is a human expression that has been fixed in tangible form. For example, a speech is not copyrightable but a transcript or recording of a speech is copyrightable. Courts have interpreted the terms "author" and "work" rather liberally, so that poems, books, songs, paintings, dance forms, and computer programs are copyrightable (A. Miller and M. Davis 2000). To be "original," a work need not be novel or nonobvious, but it must be original to the author and have some minimal degree of creativity. For example, one may copyright a reproduction of a painting if the reproduction introduces some nontrivial variations. One may copyright a compilation of facts, provided that the compilation contains some creativity in the selection, organization, or presentation of the facts (*Feist Publications v. Rural Telephone Service* 1991).

A person or organization that violates the copyrights of an original work may face civil or criminal liability. The copyright holder may sue the defendant for infringement. To prove infringement, the copyright holder must prove (a) that he or she is the author of the work; (b) that the work is original; (c) that the work is fixed in a tangible medium; and (d) that the defendant copied, displayed, performed, distributed, derived works from, or commercialized the work without permission. If the defendant intentionally violated copyrights for commercial purposes, then he or she may face criminal prosecution, which could lead to a fine or up to ten years in federal prison.

There are limitations on copyright protections. One may not copyright the facts or ideas expressed by an original work. For example, one may copyright a photograph of the moon, but one may not copyright the moon itself. One may copyright a document describing a system of accounting, but one may not copyright the system of accounting itself (*Baker v. Seldon* 1879). One may also not copyright useful inventions. For example, if a lamp contains useful and non-useful (or ornamental) parts, one may copyright the non-useful parts but not the useful ones. If one could copyright a useful work, then this would allow a private party to gain seventy or more years of control over something that should only have twenty years of protection (A. Miller and M. Davis 2000).

Another important limitation on copyright protections is the doctrine of fair use. The doctrine of fair use is a defense to an infringement lawsuit. According to this defense, one may copy, perform, display, distribute, or derive works from an original work without the consent of the copyright holder if the use is a fair use. To determine whether a use is fair, a court will consider a number of factors, including the following:

1. The nature of the use (commercial versus noncommercial, i.e., personal or academic)
2. The nature of the work (commercial versus noncommercial)
3. The amount of the use (how much was used?)
4. The significance of the part that was used (was it an important part of the work?)
5. The economic impact of the use (did it defray the economic value of the work?)

The last factor, economic impact, is often the most important one. For example, U.S. courts have found that taping a television broadcast for private viewing is a fair use, but that taping a television show for public viewing is not a fair use because public viewing of the show has potentially a much greater economic impact on the show than private viewing. The courts have also found that satire and commentary are fair uses of original works. The basic rationale for all limitations on copyrights is to strike an appropriate balance between copyrightable works and items that belong in the public domain.

6.4 Trademarks

A trademark is a symbol, a mark, a name, a phrase, or an image that a business uses to distinguish its goods or products. For example, the name "Coca-Cola," the symbol of the golden arches, and the phrase "Have it your way" are each trademarks. The owner of the trademark has the right to exclude others from using the trademark without permission for as long as he is using his trademark. To obtain trademark protection, the mark need not be novel or original but it must be distinctive. A mark is distinctive if it allows consumers to distinguish the goods of the trademark holder from other goods. Although trademarks play a very important role in business and industry, they do not have a very significant impact on scientific research. Thus, this book will not discuss them in any depth. However, it is worth noting that courts also face dilemmas relating to balancing public and private interests when they address trademark cases (A. Miller and M. Davis 2000).

6.5 Trade Secrets

Prior to the development of the patent system, craftsmen, technicians, and scientists protected their inventions through trade secrecy. One of the main reasons that governments developed the patent system was to encourage inventors and craftsmen to disclose their practical secrets so that others could learn from them. Although trade secrecy is no longer the norm in scientific and technical fields, it continues to play an important role in business and industry. A trade secret is information with economic value, which a business protects to maintain its competitive advantage (*Black's Law Dictionary* 1999). A trade secret could be a business plan, a list of clients or customers, marketing strategies, policies, recipes, or formulas. A business must take reasonable steps to maintain secrecy, such as not disclosing the protected information and requiring employees to sign confidentiality agreements. Someone who discloses (or appropriates) a trade secret without permission from the company can face civil and criminal liability and may be subject to a fine, imprisonment, or both. It is legal to use lawful means to discover trade secrets through reverse engineering or independent research. For example, a company that wanted to discover the formula for Coca Cola, one of the world's best guarded trade secrets, could conduct its own research in attempt to replicate this product. There are no limits on the length of trade secrecy protection; a business can keep a trade secret as long as it can maintain secrecy. However, trade secrets, like military secrets, are often short lived and can be difficult to protect (Foster and Shook 1993).

6.6 Intellectual Property and the Progress of Science and Technology

There are several different ways of justifying intellectual property rights (IPRs). Libertarians view property rights, including IPRs, as fundamental, moral rights similar to other moral rights, such as rights to life and liberty. The seventeenth-century British philosopher John Locke has had a great deal of influence over modern political thought concerning property. According to Locke, we can acquire property through barter or other forms of exchange or by investing our labor in something. For example, a person can acquire ownership over a flute by using his or her labor to carve the flute from a piece of wood. Under a Lockean approach, someone who writes a book should own the book because he has invested his labor in the book (Resnik 2003c).

While the Lockean approach to justifying IPRs is philosophically important and interesting, it has several flaws (Resnik 2003c). First, IPRs often do not correspond to the amount of labor invested in something. A person can invest a great deal of labor in developing a useful invention, but he will not receive a patent if the invention has been previously disclosed or is obvious. A person can invest very little effort in writing an e-mail message, yet he or she will still have copyrights over that message. The labor-mixing theory is also not a useful guide to the allocation of authorship or inventorship. For example, suppose a senior investigator spends only about five hours a week supervising a research project, while two technicians and two graduate students each spend about forty hours a week on the project. The senior investigator could still be named as an author on a paper about the project if he or she has made a significant intellectual contribution to the project, while the two technicians might not be named as authors if they have not made a significant intellectual contribution (Shamoo and Resnik 2003). Finally, it is unreasonable to think that one could acquire some properties simply by adding labor to something. For example, a person should not acquire ownership of a previously unowned tract of land simply by marking the boundaries of that land.

A slightly different libertarian approach to IPRs is based on the writing of the nineteenth-century German philosopher Georg Wilhelm Friedrich Hegel, who argued that property rights should be protected to promote self-expression, which is essential for personal development and for realizing autonomy and freedom. Under this approach, a person should have intellectual property rights over his book because he has expressed himself in the book. One problem with the Hegelian view is that society often recognizes IPRs when very little (if any) self-expression is involved. For example, copyright laws protect computer generated music, art, and animation. Trade secrecy laws protect business secrets. Another problem with the Hegelian view is that original works and inventions may result from collaborations among many different people, and are not the unique self-expression of any one person. For example, consider all the different people who contribute to the production of a motion picture.

The most influential approach to IPRs, by far, is the utilitarian view. According to utilitarians, IPRs granted by the government should promote the greatest balance of good/bad consequences for society, such the progress of science, technology, industry, and the arts (Resnik 2003c). Article 1, Section 8, Clause 8 of the U.S. Constitution (1787) provides a clear statement of utilitarian account: "Congress shall have the power . . . to promote the progress of science and the useful arts by securing to authors and inventors the exclusive right to their respective writings and discoveries." This passage draws a connection between rights granted to authors and inventors and the progress in science and the useful arts, namely, technology.

Economists have studied the relationship between intellectual property rights and scientific and technical progress. Although some studies have produced evidence that IPRs contribute to the growth of knowledge, it is very difficult to prove a causal relationship between IPRs and knowledge advancement, since it is practically impossible to conduct a controlled experiment to test this hypothesis. The best that one can hope to do is conduct a retrospective analysis of the effects of intellectual property rights on science, technology, and industry (Merges and Nelson 1990; Svatos 1996; Hall and Ziedonis 2001).

In my recent book *Owning the Genome*, I described some evidence that tends to show that intellectual property rights in the genetic sciences have led to progress in the genetic sciences (Resnik 2004c). Since the 1980s, patent applications in the genetic sciences have risen dramatically but so have publications. From 1990 to 2001, publications in the genetic sciences increased from 43,089 per year to 114,354 per year, according to the Science Citation Index. During that same period, the number of DNA patents issued per year by the USPTO increased from 265 to 2143. (See table 6.1.) Thus, there is a strong positive correlation ($r = 0.895$) between patenting and publication. Although this statistical association does not prove the patenting causes progress in the genetic sciences—there could be a confounding factor that causes patenting and publication, such as funding of R&D—the association constitutes substantial evidence in favor of the hypothesis that patenting does not undermine progress in the genetic sciences.

Table 6.1 Genetics Publications and DNA Patents, 1994–2001*

Year	Publications	Patents
1994	78,771	554
1995	87,451	603
1996	92,382	1,006
1997	98,516	1,496
1998	105,321	2,078
1999	108,650	2,066
2000	111,540	1,896
2001	118,605	2,143

*Based on Resnik (2004c).

To understand how IPRs could lead to progress in science and technology, one must consider the economic motives of scientists and research sponsors. IPRs can promote the progress of science and technology by giving incentives to inventors and private companies. IPRs allow scientists and inventors to profit from their inventions and discoveries. Idealists might argue that scientists and inventors should be motivated by the desire to benefit mankind and advance human knowledge, and, indeed, most are. But, practically speaking, most scientists and inventors also strongly respond to economic incentives. Without the incentives provided by IPRs, many scientists and inventors might refrain from engaging in R&D, or they might attempt to protect their inventions and discoveries via trade secrecy.

Of course, if a scientist or inventor seeks private funding from a company for his or her research, that company may require the scientist or inventor to seek and protect intellectual property rights. Private companies want IPRs in order to obtain a return on their R&D investments. It costs, on average, $500 million to $800 million to develop a new drug, conduct clinical trials, and bring the drug to the market. Since this process can take ten to thirteen years, a company may have only seven to ten years to recoup its investment while the drug is still under patent. Once the drug goes off patent, other companies can make generic versions of the drug without violating the patent, and the company will lose its control over the market. Moreover, only 33% of new drugs are profitable, and companies often must withdraw profitable drugs from the market due to adverse side effects or liability concerns (Resnik 2004a). Given the enormous risks involved in funding biomedical R&D, pharmaceutical and biotechnology companies require intellectual property protection to safeguard their investments. Although other industries may not take such huge risks, they still desire to protect intellectual property rights in order to profit from their R&D investments. It would not make economic sense for a company to spend millions or even billions on R&D without expecting to make a reasonable return on this investment.

Although IPRs generally promote the progress of science and technology in the long run, they can sometimes hinder progress in the short run. As we have noted many times in this book, openness is one of the most important principles of scientific research. However, intellectual property interests can interfere with the free and open exchange of scientific information. For example, prior to filing a patent application, a prudent inventor will avoid disclosing information that could jeopardize his patient, and a prudent company that sponsors his research will require her to sign a confidentiality agreement. Although the information will eventually enter the public domain via the patent application or publications, scientists will not be able to have access to this information in the short run. Furthermore, the emphasis on protecting intellectual property can have a negative impact on the academic culture, which emphasizes openness, and can encourage people to refrain from sharing items that have no apparent commercial value, such as research methods and tools (Krimsky 2003).

Let us consider the detrimental effects of a patent once it is approved. If a private company has exclusive control over an important product or process in a particular domain of inquiry, then the company could deter scientific and technical progress in that domain by refusing to license others to use or commercialize the product or process or by charging high licensing fees. For example, geneticists, clinicians, and patients have complained about the price of Myriad Genetics licensing fees for its BRCA1 and BRCA2 genetic tests for hereditary breast cancer (Resnik 2003a). Finally, as noted earlier, difficulties with negotiating licenses, as well as transaction costs and licensing fees, could constitute a heavy toll on downstream research (Heller and Eisenberg 1998). Thus, the private control of scientific and technical information can have negative effects on the progress of science and technology and the ethos of science (Demaine and Fellmeth 2002, 2003).

Some writers have suggested that one way of dealing with these licensing issues would be to form a patent pool for biotechnology. A patent pool would be an independent organization that manages patents and issues licenses. Different companies could place their inventions in a pool. A patent pool would drastically reduce transaction costs related to licensing, since people who want to license products or processes could negotiate with the organization operating the pool rather than a dozen or more different patent holders (Resnik 2003g; Grassler and Capria 2003). However, a patent pool might not be very successful if companies think that they can make a better profit by staying out of the pool. For example, a company with a large number of patents or a company with a highly prized patent might stay out of a patent pool.

6.7 Balancing Public and Private Interests

Since IPRs can have positive as well as negative effects on the advancement of knowledge, the best policy for science and society would be one that strikes an appropriate balance between public and private control of information. Although private parties should be allowed to control some information for limited periods of time, a great deal of information should be freely available to the public. The important policy issues for government agencies, the courts, and legislators will revolve around finding that appropriate balance of public and private control (Resnik 2004c). The following are some areas where the legal system can set and adjust this balance.

Patents: Subject matter. Deciding what counts a patentable subject matter provides policy makers with a valuable opportunity to balance private and public interests related to IPRs. As noted earlier, the U.S. legal system does not allow patents on items that are not regarded as belonging to a patentable subject matter, such as laws of nature, natural phenomena, and mathematical formulas. To be patentable, an invention must be a product of human ingenuity, not a product of nature. For example, the USPTO grants patents on isolated and purified chemicals that occur in living things, such as proteins, DNA, and hormones, but it does not grant patents on these chemicals in their natural state (Doll 1998). What is the difference between DNA as it occurs in nature and isolated and purified DNA produced under laboratory conditions? The difference, according to the U.S. PTO, is that human ingenuity is required to produce an isolated and purified sample of DNA, but no human ingenuity is required to produce DNA as it occurs in nature (Resnik 2004c). Human ingenuity is also required to isolate and purify water, but water is not patentable. Perhaps the difference between purified water and purified DNA is that more human ingenuity is required to isolate and purify water. But how much human ingenuity is enough ingenuity? What does it take to transform a product of nature into a product of human ingenuity? Some have suggested that the transformation must be substantial (Demaine and Hellmeth 2002, 2003). But what is a substantial transformation?

One encounters similar problems with other difficult cases in patent law, such as computer programs. The U.S. Supreme Court has held that the mathematical algorithms and formulas contained in computer programs are not patentable but that programs may be patentable (*Diamond v. Diehr* 1981). Even though parts of programs are not patentable, programs, as a whole, may be patentable because they are human-invented processes that perform useful operations, such as curing rubber or manufacturing automobiles. But how much human ingenuity is required to transform a set of equations or algorithms into a useful process? At some point, the courts must draw a line between algorithms/equations, which are abstract ideas, and computer programs, which are practical applications.

Unless one can come up with a widely accepted metaphysical theory that distinguishes between natural and artificial things and abstract ideas and practical applications, deciding whether a product or process should be classified as belonging to a patentable subject matter comes down to pragmatics. To determine whether a product or a process should be considered to be patentable, one must ask whether classifying the product or process as patentable serves the goals and purposes of the patent system (Demaine and Hellmeth 2002). How would treating a product or process as patentable affect the progress of science and technology? One should also consider whether regarding a product or process as patentable would have any bearing on moral values, such as respect for human dignity and rights, respect for nature, and social justice.[1] In answering these questions, patent agencies, judges, and legislators should balance public and private interests.

Patents: Scope. Deciding the scope of patents also affords policy makers with an opportunity to balance public and private interests. If the scope of a patent is too broad, then the patent holder will have too much control over the market. The patent holder will be able to block competitors from entering the market, which will result in less scientific and technical innovation as well as in market failures. If the scope of a patent is too narrow, then the patent holder will not have enough control over the market, which will provide insufficient incentives for invention and investment (Jaffe and Lerner 2004).

The research exemption. The research exemption is another area where the legal system could balance public and private interests. As noted earlier, this narrow exemption is now on shaky ground as a result of the ruling in *Madey*. Academic researchers no longer enjoy a blanket exemption from patent infringement; they can only take advantage of this exemption for if their work is noncommercial. The legal problems and issues resulting from *Madey* have tipped the IPR scale toward private interests. To restore this delicate balance, governments should consider amending the patent laws to strengthen and clarify the research exemption (Eisenberg 2003; Resnik 2004c). Academic (or nonacademic) researchers should be exempt from patent infringement if their work is conducted for noncommercial purposes and it will not significantly impact the economic value of the patented invention.

Copyrights: Subject matter. The legal system can also balance public and private interests when determining whether something is copyrightable. One of the most important cases relating to the subject matter of copyrights in the United States is *Feist Publications Inc. v. Rural Telephone Service Company* (1991). In this case, the U.S. Supreme Court ruled that Rural Telephone Company had no copyright protection over its phonebook because it did not meet the legal requirements for a compilation under the Copyright Act. The court held that the phonebook was not copyrightable because it did not exhibit a minimum level of creativity in the selection or organization of the data. In the wake of *Feist*, copyright protection for databases has been uncertain. Since databases play a very important role in scientific research, it is important to balance public and private interests with relating to the databases. On the one hand, it is important to place a great deal of information in the public domain. On the other hand, it is also important to provide private database creators with adequate economic incentives. The European Union has adopted laws that protect noncreative databases for fifteen years. Creative databases are protected under copyright laws. One might argue that the best way for the United States to balance public and private interests is to follow the European model regarding databases and to include a fair use exemption (Resnik 2003f).

Copyrights: Fair use. The doctrine of fair use is another area where the legal system balances public and private interests. If fair use is too narrow, then the private interest is too strong and the public interest will be too weak. If fair use is too broad, the private interest will be too weak and the public interest will be too strong. Fair use should be broad enough to allow some unauthorized uses that do not undermine the commercial value of an original work, but not so broad that it reduces economic incentive for authors and businesses. In the United States, the legislature, government agencies, and courts have taken steps allow noncommercial activities relating to education or research to be included under the doctrine of fair use. These different bodies within the legal system should continue to clarify the doctrine of fair use in response to changes in communication and information technology.

Enforcement. It is also worth mentioning that the enforcement of intellectual property laws plays a key role in how the legal system balances public and private interests. If the laws are not well enforced, private parties may attempt to make unfounded or extravagant claims to intellectual property. This is especially true in patenting, where USPTO has awarded patents that were later proven invalid in court. Companies have attempted to patent the same products twice, to patent inventions without a clear practical use, or to obtain a patent that was excessively broad in scope. Since companies may try to abuse or game the patent system, it is important for governments to make sure that the laws are strictly enforced, and that the government does not give away IPRs unfairly (Jaffe and Lerner 2004).

6.8 Conclusion

IPRs can stimulate the progress of science and technology by providing incentives for researchers and private companies, but they can also impede the progress of science and technology by undermining the ethic of openness. There are extremists on both sides of the debate about IPRs. Those who take a Marxist/socialist view of IPRs would like to eliminate them completely (Martin 1995). Those who take a strong libertarian view of IPRs would like to strengthen and extend them. Prudence suggests that the optimal solution lies somewhere between these two extreme views; society should develop IPR polices that strike an appropriate balance between public and private interests.

In chapter 1, I noted that Krimsky claims that "privatization of knowledge has replaced communitarian values" (2003, 7). We can now see that this statement is an exaggeration. At no time in the history of science have researchers accepted communitarian values to the exclusion of non-communitarian ones. As we saw in chapter 4, secrecy, or the desire to hoard scientific knowledge, and openness, or the desire to share scientific knowledge, have always coexisted in science. Although science has become more privatized in the last two decades, the legal system has always balanced public and private interests relating to IPRs and will continue to do so. The key is to strike the best balance of public and private interests in IPR law and policy. Since changes in technology, such as computers and the Internet, can affect this balance, the executive, legislative, and judicial branches of government should continually revise polices, laws, and judicial interpretations in response to changes in the way the information is created, transmitted, and controlled.

SEVEN

PUBLICATION

Openness and Accountability

The pen is mightier than the sword.

—Baron Edward Bulwer-Lytton

We have noted how financial interests in research can lead to numerous ethical problems with publication and authorship, such as suppressed publication, duplicate publication, ghost authorship, and honorary authorship. This chapter will reexamine these problems and propose some solutions for addressing them.

7.1 Publication Bias

In chapter 1, we considered several cases where private companies sought to block the publication of negative research data and results. Boots Pharmaceuticals tried to prevent Betty Dong from publishing her research demonstrating that Synthroid is not better than competing drugs, and Apotex tried to stop Nancy Olivieri from publishing her work concerning the dangers of deferiprone. The researchers in both of these cases worked for academic institutions. The researchers had both signed confidential disclosure agreements (CDAs) with the companies as well.[1] We also noted in chapter 1 that many researchers sign confidentiality agreements when they receive contracts or grants to conduct research for private companies.

We also have discussed, in chapters 1 and 4, the strong link between the source of funding and research results. Published research studies almost always favor the sponsor's products. One explanation for this bias is that companies often do not publish negative results.

Companies have no legal obligation to publish the research that they conduct. If a company is seeking FDA approval of a new drug, biologic, or medical device, then it must submit all its data to the FDA. The FDA will review data from the company and data that may be available from other sources. However, the FDA treats that data as a trade secret and does not require the company to publish it. As a result of this tendency to publish positive results but not negative ones, the research record can be skewed in favor of the new drug. Researchers or clinicians who want information about the drug will not have access through the research record to these unpublished, negative findings. The unrevealed risk of using anti-depressants to treat depression in children, discussed in chapter 1, illustrates this problem. The medical community did not have access to important information about teenage suicide rates because the data had not been published.

Dong and Olivieri sought to publish negative results, despite harassment and legal action from the private companies that sponsored their work. There is evidence, discussed in chapter 1, that many researchers have delayed publication for various reasons and that many researchers have signed CDAs that restrict publication. Thus, many other researchers have probably declined to publish negative findings out of fear of harassment or legal liability. However, this issue requires further study.

Scientific journals also play a role. There is strong evidence of a bias toward positive findings in published research (Simes 1986; Easterbrook et al. 1991; Callaham et al. 1998). One simple explanation for this bias would be that editors and reviewers may prefer to publish positive results. However, there is some evidence that refutes this hypothesis. One study found that among submitted manuscripts, there is no statistically significant difference in the publication rates of positive and negative results (Olson et al. 2002). According to the authors of this study, the best explanation for the phenomena of bias toward positive results is that authors are less likely to submit manuscripts about negative findings to journals. Another reasonable explanation of this bias is that it is more difficult to prove a negative finding with a high degree of statistical significance, in other words, P-value 0.05 or less, than it is to prove positive findings, and journals prefer to publish results with low P-values (Resnik 2000b).

Researchers can usually publish negative findings, but they may need to seek publication in lesser-known, lower-impact journals. The trend of publishing positive findings can also have a significant impact on research studies that assess the published research, such as meta-analyses and literature reviews. A study of the research record is only as good as the record itself; if the record is biased toward positive results, then a study of the record will be biased as well. Clearly, the tendency to only publish positive findings undermines two important scientific norms: objectivity and openness. How should the scientific community respond to this problem?

One potential way of dealing with this problem is to develop ways of publishing negative results and making them available to scientists. One such proposal would be a clinical trial registry, discussed in chapter 1 (Simes 1986; Rennie 1999). The purpose of a clinical trial registry is to make published and unpublished data from clinical trials available to researchers and clinicians, who can analyze and interpret the data to better understand benefits, risks, adverse events, and other aspects of a new drug, biologic, or medical device. One example of a clinical trial registry is the Australian Clinical Trails Registry (ACTR). The aims of the registry are to "provide a general description of clinical trials research, identify clinical trials planned and in progress, enable protocol details to be registered before the results are known, provide a means for selecting trials for meta-analysis and review articles, ensure that trials are ethical and worthwhile, in the long term, improve clinical care and clinical practice" (ACTR 2005). Another example of a registry is the International Standard Randomized Controlled Trial Number (ISRCTN). According to the ISRCTN:

> Without these innovative tools [the registry], clinicians, researchers, patients and the public will remain in ignorance about ongoing and unpublished trials or confused about which trial is which. Opportunities for collaboration and reducing duplication of research effort will be missed. Publication bias and undeclared over-reporting will lead to misleading conclusions being drawn about the forms of care most likely to benefit patients. Patients may even be subjected to trials seeking evidence that is already available. Not reporting RCTs is increasingly seen as scientific and ethical misconduct, and the pressure to register trials to reduce biased under-reporting is growing. (ISRCTN 2003)

Although Congress has considered some legislation that would mandate clinical trial registration, clinical researchers in the United States and most other countries are not required by law to register trials or publish negative results (Couzin 2004). Many clinical trials in the United States are registered at ClinicalTrials.gov, a clinical trial registry sponsored by the NIH and the National Library of Medicine (ClinicalTrials.gov 2006). ClinicalTrials.gov includes over 27,000 studies sponsored by the NIH and private industry.

Even though many companies appear to be responding to the voluntary registration system, given the scientific and clinical significance of the problem of publication bias, mandatory registration of clinical trials would be a wise idea. As mentioned in chapter 1, the International Committee of Medical Journal Editors (ICMJE) now requires that researchers register clinical trials as a condition for publishing the results of those trials in ICMJE journals. It remains to be seen whether the U.S. government or other governments take legal action to require clinical trial registration.

Although the need for a database of unpublished results is most acute in the research that involves human subjects, other scientific disciplines can benefit from these databases as well. For example, the National Center for Biotechnology Information (NCBI) sponsors a variety of public databases, such as GenBank, a depository for DNA sequence data (NCBI 2003). Many scientists in genetics, molecular biology, genomics, and biotechnology deposit information in this database, and all also use this information.

Numerous granting agencies and journals require scientists to make their data available through public databases (Rowen et al. 2001). However, private companies may not always agree with this policy. As we saw in chapter 1, Celera Genomics wanted to make its data available only through its own, private database. It is worth noting that physicists, chemists, geologists, and astronomers have also developed public databases accessible to researchers.

While databases for unpublished results can help to increase access to data and overcome publication biases, these databases also pose some hazards. One problem with making unpublished results/data available to researchers is that this may bypass peer review, which is essential for ensuring the quality of scientific research. Published research data are subjected to peer review, but unpublished data are not. The cold fusion case, discussed in chapter 1, illustrates some of the problems with non-peer reviewed publication. Media coverage of research presents a similar concern. Frequently, the media provides extensive coverage of abstracts presented at scientific meetings but never published (Schwartz et al. 2002). The NIH's controversial E-biomed proposal would also have bypassed normal peer review (Relman 1999). In one version of this proposal, researchers could post their data and results to the NIH's Web site after their work was approved by two reviewers selected by a governing board. Although the data and results would be peer reviewed, one might question the qualifications of the reviews and editors working for the governing board. Normally, peer-reviewed papers are evaluated by editors or reviewers working for scientific journals or societies with a specific disciplinary focus or area of expertise. One additional problem with bypassing peer review is that a meta-analysis that includes unpublished studies may not be as reliable as one that includes only published studies, since the meta-analysis may be based on unreliable data. The computer science axiom "Garbage in, garbage out" applies here.

How should the scientific community deal with the problem of bypassing peer review channels? First, one could establish governing boards for databases that review all submissions to the databases. Submissions would not be posted until they pass some sort of peer review. This is essentially the E-biomed proposal. The problem with this idea is that the reviewers and editors might not be well qualified to review the submission to the database. A second proposal would encourage scientific journals to create electronic repositories for data and results that are not published in the full version of the print or electronic journal. Submissions to the repositories would be peer reviewed by the editors and reviewers who work for the journal, according to the usual methods and procedures. Data/results published in the repositories would be fully indexed and abstracted as if they were published in journal. Essentially, this proposal would allow for another type of publication. The editors could decide that some papers merit publication in the journal, while other papers, though publishable, only merit publication in the repository. This would increase the quantity of data and results available to research, and help to overcome the problem of publication bias, without bypassing peer review.

Another potential solution to the problem of bias toward positive results is for universities to carefully scrutinize the types of contracts that they make with private corporations that sponsor research (Bok 2003; Krimsky 2003). These contracts should attempt to balance commercial and academic interests. On the one hand, companies should not be allowed to suppress research conducted at academic institutions. On the other hand, companies should have the right to at least review data and results prior to publication, in order to ensure the accuracy of the work and to secure their intellectual property interests. So, companies should be allowed to delay the publication of data/results for a reasonable time. What kind of delay would be unreasonable? Any number we choose here would be somewhat arbitrary, but a delay of more than six months would be unreasonable, barring extraordinary circumstances. Universities should not sign agreements that allow companies to suppress or unreasonably delay publication of data or results.

One obvious shortcoming of this proposal is that some companies may decide to use contract research organizations (CROs) to conduct clinical trials if academic institutions place restrictions on research contracts that the companies regard as burdensome or restrictive. As noted in chapter 1, CROs are taking an increasingly large bite out of the clinical trial market. Leaders of academic institutions may be wary of placing restrictions on contracts with private companies because they may fear that they will lose even more of their business to CROs. While most academic researchers want to publish their work and endorse the ethic of openness, physicians in private practice, who are also recruiting patients for clinical trials, may have no qualms about allowing research sponsors to suppress or delay publication of data or results.

This economic quandary epitomizes the central ethical dilemma of academic biomedical research: if researchers want to adhere to the highest standards of ethics, they may fail to capitalize on the financial opportunities presented by clinical trials; if researchers want to capitalize on these financial opportunities, they may have to compromise their ethical standards. Ideally, ethics should take priority over economic interests in the conduct of clinical trials (and other types of research). Nevertheless, one can see how university and hospital administrators might reach a different conclusion.

7.2 Duplicate Publication

Duplicate publication, which was discussed briefly in chapter 4, has occurred for decades. Duplicate publication occurs when scientists publish essentially the same data or results in different journals, Twenty-five years ago, Huth (1986) reproached researchers for engaging in this practice. For the past two decades, various scholars and professional organizations have condemned this practice as deceptive and wasteful (National Academy of Sciences 1992; LaFollette 1992; ICMJE 2005). Estimates of the incidence of duplicative publication range from 0.017 to 306 repetitive pairs of papers per 1,000 papers (Huth 2000). The wide variation in these estimates is probably due to differences in defining the concept as well as differences in gathering data. A recent notice of duplicate publication published in a prominent medical journal indicates that the phenomena occurs even in the top journals and has important implications for the ethics of science (Gregory, Morrissey, and Drazen 2003).

Although there are no empirical studies that address the financial aspects of duplicate publication, one may hypothesize that financial interests play some role in causing this phenomena. First, authors may have a financial motive to engage in this practice in order to increase the quantity of their publications, which is an important factor in tenure and promotion. Very simply, researchers may engage in duplicate publication to keep their job or win a promotion (Shamoo and Resnik 2003). Second, research sponsors may also encourage this practice in order to publish more positive results and skew the published record (Angell 2004). For example, if a pharmaceutical company can claim that six published studies support its product, this appears to be much more impressive than two published studies. A meta-analysis of the research record that included duplicate studies of the same, positive results would be skewed.

Duplicate publication undermines the ethics of science in several ways. First, it is dishonest and deceptive. Second, it allocates credit unfairly by giving researchers more credit for their research or accomplishments than they deserve. Third, it can threaten the objectivity of the research record by making a particular hypothesis appear to have more empirical support than it actually has. Fourth, it is wasteful. Publishing the same results in two different journals taxes the resources of journals and peer reviewers. Thus, duplicate publication is a serious problem for the ethics of research. How should the scientific community respond to this problem?

First, many journals already have policies against this practice. Journals that have not adopted policies prohibiting duplicate publication should follow the example of the ICMJE (2005), which has taken a strong stance against duplicate publication for many years. Second, universities and professional organizations should also adopt policies that prohibit duplicate publication. Third, although policies are important, some type of enforcement of the rules may also be required. It is now possible to use computer programs to compare documents to determine their degree of similarity. University professors have used computer programs to detect plagiarism since the mid-1990s. One could apply these methods to the research record as well. Journals could make a quick check of an article when it is submitted and again just prior to publication. They could use computer programs that search available databases to detect duplication. If a program indicates possible duplication, the editors could examine the two articles, and contact the authors, if necessary. If an editor catches an author attempting to practice duplicate publication, he or she could take disciplinary action against the researcher, such as contacting the researcher's institution or other journal editors.

7.3 Salami Science

A problem related to duplicate publication is dividing a scientific research project into a dozen or so small publications instead of two or three larger publications, without a legitimate reason. (This problem was also mentioned briefly in chapter 4.) Sometimes researchers have legitimate reasons for dividing up their projects into different publications, due to the complexity, volume, and diversity of their data and methods. However, sometimes researchers subdivide their projects with a legitimate reason. This practice, also known as publishing according to the least publishable unit (LPU) or "salami science," was first described by Huth (1986). LaFollette (1992) has also documented this problem. The reasons that a scientist might practice salami science are fairly straightforward: increasing the number of publications per project increases one's total publications and therefore helps to advance one's career. A company might also practice salami science in order to increase the number of favorable publication regarding one of its products. Thus, money can play a factor in encouraging salami science. Salami science, like duplicate publication, is wasteful, can skew the publication record, and misallocates credit.

To minimize salami science, journals, research institutions, and professional associations should develop and enforce policies that discourage unnecessarily dividing up research projects into smaller publications. These policies should describe some legitimate reasons for publishing different parts of a research project in different journals. Also, researchers should educate students, fellows, and trainees about these issues.

7.4 Electronic Publication

In an effort to promote the rapid sharing of data, methods, ideas, and results in science, many researchers have experimented with various forms of electronic publication, including publication on personal Web pages, publication in paper and electronic media, and electronic journals. Electronic publication offers many benefits to science but also has several pitfalls (McLellan 2000). The benefits of electronic publication are that it can greatly reduce the cost of publication and increase the speed and quantity of publication (Butler 1999). Web pages and electronic journals do not have the same type of space constraints as print journals. It usually takes less time to update a Web page or electronic journal than to publish a print journal. It also usually costs much less money to publish electronically. The pitfalls of electronic publishing are that some types of electronic publication, such as publication on personal Web pages, blogs, or the E-biomed proposal (discussed earlier), may evade traditional peer review. Electronic journals also may not have the same prestige as paper journals, although this will soon change.

Although electronic journals are becoming increasingly common, and most journals now publish electronic and paper versions, there remain some obstacles to electronic publication. Some of these, such as the problems with peer review (discussed earlier), can be handled provided that journals take steps to establish conflict of interest policies and ensure that there is adequate peer review by qualified reviewers. Another problem—the problem of paying for electronic journals—may be more difficult to solve. Traditional scientific journals have been funded by fees from subscribers, fees from memberships in scientific organizations, or, in some cases, money from sponsoring organizations, such as the National Cancer Institute or the National Institute for Environmental Health Sciences. The costs of managing, editing, and publishing scientific journals have risen dramatically in the last two decades. One reason that electronic publishing has become more common is that scientists are looking for ways to save money on the cost of publication. Although electronic journals are generally less expensive than print journals, they are not cheap. Scientists must usually still pay for access to electronic versions of articles through subscription fees or other fees.

One way of dramatically increasing openness and sharing in science would be to make electronic journal articles available for free. Open access publishing would do just this. Any scientist in the world can download articles from an open access journal for free. The cost of the journal is paid up front by scientists who submit papers, by sponsoring organizations, or by both (Malakoff 2003). The NIH and the Wellcome Trust, a U.K.-based research foundation, have backed the idea of open-access publishing (Wadman 2004; Clery 2004).

Even though open access publishing can increase the flow of information in science, it is likely to create financial problems for electronic journals. One open access journal, *PLoS Biology*, published by the Public Library of Science, plans to charge authors $1500 per published paper. Other journals would need to charge far more than

$1500 to authors if they went to an open access system. For instance, *Science* estimated that it would need to charge about $10,000 per published paper (Malakoff 2003). Even the marginal fee of $1500 could pose a significant economic barrier to some authors who are seeking publication, and it would at least pose an economic deterrent to others. Although free and open access to scientific data, results, and papers sounds like a good idea, there is no getting around the practical problem that someone must pay for the costs of publishing and sharing research, whether it is the publisher, author, or reader/user. This is not a case of greed interfering with sharing in science; it is simply a matter of the economics of publishing. There is no free publication.

Though well intended, a policy of requiring funded researchers to publish only in open access journals could have adverse effects. In most fields, where an article is published matters a great deal. Researchers want their results to reach the appropriate audience and to have an impact on other scientists. Suppose that the optimal journal for a particular article is not an open access journal but that researchers are required to publish only in open access journals. If this situation occurs, researchers may be forced to publish articles in less-than-optimal venues. If this becomes an ongoing problem, researchers may decide to pursue funding from organizations that do not mandate open access publishing.

Perhaps a workable compromise would be for journals to experiment with different types of open access, and for funding agencies to experiment with different types of open access policies. All journals already make abstracts available via open access. They could also experiment with different types of open access. A journal could publish some, but not all, articles on an open access basis. The *New York Times*, for example, publishes its articles on an open access basis for the first few days, and then charges a fee to access archived articles (articles that are older than a few days). Scientific journals could experiment with a policy like this one or one very different from it, such as free access to archived articles but not to recent ones. Likewise, funding agencies could require that some, but not all, publications would be available on an open access basis.

7.5 Access to Data Following Publication

The ethic of openness, which plays such an important role in publication, requires researchers to allow scientists to have access to the data supporting a manuscript following publication. Since journals have limited space, when researchers publish a scientific paper in a journal, they often do not include all the data that one would need to reproduce the results. They also do not include protocols and standard operation procedures, as well as materials used in research that are physically impossible to publish, such as reagents, cell lines, and laboratory equipment. In order to promote the replication of results, scientific criticism, and collaboration, it is important for researchers who are not involved in a publication to have access to all of the data and materials that support it. Many journals and funding agencies have instituted data-sharing policies. The NIH requires intramural and extramural researchers to share data and research materials with other researchers following publication (NIH 2002). Many journals require researchers to deposit DNA sequence data in a publicly accessible database, such as GenBank. The journals *Science* and *Nature* require researchers to make publishable items available to other researchers through their Web sites as a condition of publication.

Even though many scientists and scientific organizations support data sharing following publication, some resist it. As we saw in chapter 1, Celera wanted to place some restrictions on access to its DNA sequences data that supported the publication of its paper on the human genome. Although the company allowed researchers to access the data for academic purposes, it wanted to protect its investment from people who might exploit the database for commercial gain. This concern—that someone else might exploit hard-won data—is not unique to industry. Academic researchers have also had qualms about making all their data available to outsiders, since they want to be able to continue to develop additional publications from the data. On the one hand, one might argue that researchers have a right to analyze the data that they have generated, and that they should be able to exclude other researchers from analyzing it until they have finished using it for their own publications. Indeed, some researchers might delay publication of their first paper from a database until they ready to publish several papers at once, in order to avoid being scooped. But on the other hand, one might argue that researchers should not hoard their data once they have generated it and published it. All science is built, to a certain extent, on preexisting data and information. There are thousands of public databases that anyone could analyze at any time. While researchers should be allowed to control the use of their data prior to publication, they relinquish this control when they publish because other researchers need to have access to this data to replicate experiments, critically examine the research, and so on. A researcher who generates data gets the first crack at analyzing the data, but once he produces a publication using that data, other researchers should be able to analyze and derive publications from his data.

The best policy lies somewhere between these two extremes (Rowen et al. 2001). The problem of sharing databases was not a significant issue many years ago because research was different. Two decades ago, research was hypothesis-driven instead of data-driven. Scientists would gather data to test a specific hypothesis or theory and then publish the data in an article or presentation assessing that hypothesis or theory. Today, scientists often develop large databases that can be used to test many different hypotheses or theories and can lead to dozens or even hundreds of publications. If a database is likely to generate one hundred publications, it does not seem fair that the person who develops the databases should garner only one publication from it, while other people, who did not develop the database, claim the other ninety-nine publications. A fair policy would allow the researcher to develop a limited number of publications from the database, for a limited time, before others could publish from it. Journals, granting agencies, and institutions should develop policies to reflect this approach.

There are also some financial and practical problems related to data sharing. It takes time, money, and human resources to share data. Some researchers may avoid sharing because they lack sufficient resources or they regard sharing as an inconvenience that interferes with research (Campbell et al. 2002). To deal with these problems, research institutions and funding agencies should provide researchers with the resources they need to share data. Each institution should store its own data, but it should not have to pay to transfer data to another institution. Institutions requesting data should also bear some of the costs of transfer, such as shipping or copying fees.

The sharing of research materials, such as reagents, cell lines, tissues, and transgenic animals, poses a different set of problems because research materials may have proprietary restrictions and may be more difficult to share. For example, a scientist may conduct research on a patented cell line by obtaining a license from the patent owner. The license may allow the research to publish and share data derived from the cell line, but it may also prohibit the researcher from transferring the cell line to others without permission. Researchers may encounter a number of practical difficulties in sharing research materials that are dangerous, such as pathogens or radioactive isotopes; delicate, such as cell lines or some types of reagents; or large, such as fossil collections, artifacts, and so on. Additionally, the cost and difficulty of storing and transferring research materials may have an adverse impact on researchers' willingness to share. While the ethic of openness still applies to the sharing of research materials, practical and proprietary restrictions may, understandably, hamper sharing. In some cases, it may be desirable for researchers to enlist the services of private companies to handle the logistics of sharing research materials. A private company could store and maintain materials for a researcher, who can then refer requests for sharing to the company, and the materials would be available to the public for a reasonable fee.

7.6 Misrepresentation of Authorship

Chapter 4 examined some problems with authorship. It also mentioned that financial interests probably help to encourage these unethical practices. People may have financial reasons for failing to give credit where it is due, as well as financial reasons for giving credit where it is not due. One survey indicates that authorship disputes are becoming more frequent (Wilcox 1998). Chapter 4 also discussed some policies that journals have adopted to prevent these problems. This chapter will discuss some other proposals for dealing with authorship problems in science.

In the mid-1990s, several authors proposed that journals develop policies that define the roles of authors more clearly, on the grounds that these policies would promote accountability, responsibility, and honesty in authorship (Pennock 1996; Resnik1997; Rennie, Yank, and Emanuel 1997). Since the year 2000, several journals, including, *Science*, *Nature*, and the *Journal of the American Medical Association* (*JAMA*), have experimented with requiring authors to describe their contributions in the manuscript. Contributions of authors typically include responsibilities like acquiring data, designing the experiments, analyzing and interpreting data, drafting or editing the manuscript, and technical or administrative support. According to the policy adopted by *JAMA*:

> Each author should have participated sufficiently in the work to take public responsibility for appropriate portions of the content. One or

> more authors should take responsibility for the integrity of the work as a whole, from inception to published article. Authorship credit should be based only on (1) substantial contributions to conception and design, or acquisition of data, or analysis and interpretation of data; and (2) drafting the article or revising it critically for important intellectual content; and (3) final approval of the version to be published. Conditions 1, 2, and 3 must all be met. Authors are required to identify their contributions to the work described in the manuscript. With the cover letter include the authorship form with statements on (1) authorship responsibility, criteria, and contributions, (2) financial disclosure, and (3) either copyright transfer or federal employment. Each of these 3 statements must be read and signed by all authors. (*JAMA* 2003)

Other journals should follow *JAMA*'s example because doing so will promote accountability, responsibility, and honesty in authorship.

Although scientific journals can promote ethical publication practices, it is also important for research institutions to help promote this worthwhile goal. Research institutions should also consider adopting policies that prohibit misrepresentation in authorship. Unethical authorship practices can be viewed as a violation of the norm of honesty (Shamoo and Resnik 2003). A manuscript that unethically excludes or includes an author misrepresents the authorship of the paper. Although this type of deception does not have the same consequences for science as fabrication or falsification of data, it should still be taken seriously. Excluding someone from receiving credit for authorship may even be considered plagiarism, if they person who was excluded made a substantial contribution to the paper with no acknowledgement whatsoever. Research institutions should make it clear that misrepresenting authorship is a violation of ethical standards. These policies should prohibit plagiarism, as well as ghost authorship and honorary authorship. Although some institutions have authorship policies, most do not (Wilcox 1998). A good example of an institutional policy is the one adopted by Harvard University's Medical School:

> 1. Everyone who is listed as an author should have made a substantial, direct, intellectual contribution to the work. For example (in the case of a research report) they should have contributed to the conception, design, analysis and/or interpretation of data. Honorary or guest authorship is not acceptable. Acquisition of funding and provision of technical services, patients, or materials, while they may be essential to

the work, are not in themselves sufficient contributions to justify authorship.

2. Everyone who has made substantial intellectual contributions to the work should be an author. Everyone who has made other substantial contributions should be acknowledged.
3. When research is done by teams whose members are highly specialized, individual's contributions and responsibility may be limited to specific aspects of the work.
4. All authors should participate in writing the manuscript by reviewing drafts and approving the final version.
5. One author should take primary responsibility for the work as a whole even if he or she does not have an in-depth understanding of every part of the work.
6. This primary author should assure that all authors meet basic standards for authorship and should prepare a concise, written description of their contributions to the work, which has been approved by all authors. This record should remain with the sponsoring department. (Harvard University Medical School 1999)

The NIH Guidelines for the Conduct of Research in the Intramural Programs include a useful policy on authorship:

> Authorship is also the primary mechanism for determining the allocation of credit for scientific advances and thus the primary basis for assessing a scientist's contributions to developing new knowledge. As such, it potentially conveys great benefit, as well as responsibility. For each individual the privilege of authorship should be based on a significant contribution to the conceptualization, design, execution, and/or interpretation of the research study, as well as a willingness to assume responsibility for the study. Individuals who do not meet these criteria but who have assisted the research by their encouragement and advice or by providing space, financial support, reagents, occasional analyses or patient material should be acknowledged in the text but not be authors. Because of the variation in detailed practices among disciplines, no universal set of standards can easily be formulated. It is expected, however, that each research group and Laboratory or Branch will freely discuss and resolve questions of authorship before and during the course of a study. Further, each author should review fully material that is to be presented in public forums or submitted (originally or in revision) for publication. Each author should be willing to support the general conclusions of the study. (NIH 1992)

7.6 Conclusion

Economic forces and financial interests can affect publication practices in science. Money plays a major role in many of today's ethical concerns in scientific publication, such as bias, duplicate publication, salami science, access to data and materials, and undeserved authorship. To address these problems and issues, scientific journals, research institutions, professional associations, and funding agencies should develop polices and guidelines for scientific publication. The ICJME and the Committee on Publication Ethics, which have been leaders among scientific journals, should continue to examine publication practices and promote ethical standards in publication. Journals should follow the standards set by these two organizations. Granting agencies, such as the NIH and the NSF, should continue to support research, education, and policy development pertaining to ethics and integrity in scientific publication. Universities, colleges, and government and private laboratories should sponsor educational and training programs that address the ethics of scientific publication.

EIGHT

GOVERNMENT FUNDING OF R&D

Science as a Public Good

That government is best which governs least.

—Henry David Thoreau, *Civil Disobedience*

8.1 Introduction

Chapters 5, 6, and 7 focused on ethical problems related to industry sponsorship of research and the financial interests of researchers and research institutions. One might be tempted to conclude from this discussion that today's ethical problems and concerns in scientific research all stem from private funding and financial interests. If one could eliminate private money and private interests from science, then science would be ethically pure.

Nothing could be further from the truth. Ethical problems can also occur when the government sponsors research for the benefit of the public. Public sponsorship of research creates a different set of ethical problems; it does not eliminate all ethical problems. Moreover, many of the ethical problems that can occur in science, such as misconduct, wasteful publication, inappropriate authorship, taking advantage of students, abuse of the peer review system, and violations of the rights and welfare of research subjects, can occur regardless of whether science is sponsored by the government or by private industry, is pursued for the sake of profit or for the sake of truth, or takes place in private laboratories or in academic institutions.

Several recent books and articles—for example, Angell (2004), Press and Washburn (2000), Bok (2003), Goozner (2004), Krimsky (2003)—focus on the ethical problems resulting from private industry's relationship to science. Clearly, there are many reasons, most of which this book has explored, to be concerned about the relationship between science and private industry. However, it would be unwise and unfair to shine the critical spotlight only on private science. This chapter will seek to rebalance our ethical assessment of science and address some problems that can arise when the government sponsors science. Like the previous chapters, this chapter will also consider how money can create or exacerbate various ethical problems.

8.2 How the U.S. Government Supports Research

This chapter will describe and explain how the U.S. government sponsors research. Although other industrialized nations have somewhat different government structures and procedures, some of the basic problems and tensions examined in this chapter also apply to other industrialized nations.

As noted in chapter 1, U.S.-government support of research began to rise steadily during World War II. A very influential rationale for government support of research was developed by President Truman's science advisor, Vannevar Bush, a physicist who is distantly related to U.S. presidents George H. Bush and George H. W. Bush. In *Science: The Endless Frontier* (1945), V. Bush argued for strong government support of basic research on the grounds that basic research builds a general knowledge base that can be used by other sciences and applied disciplines, such as medicine, engineering, and agriculture. Thus, investment in basic research leads to practical applications. Another argument V. Bush developed for government investment in basic research is that this would develop a scientifically trained labor force, which would play a role in applying scientific knowledge in the development of technology and the solution to practical problems. Bush realized that it was also important to invest in applied research, but he contended that the government did not need to devote as much money to applied research, since private companies would be interested in sponsoring research with commercial or practical value, and private companies would not be interested in sponsoring basic research. To use an economic analogy: basic science is a public good (Shamoo and Resnik 2003). Finally, V. Bush also argued that scientists should be self-regulating. Scientists, not the government, should decide whether research should be conducted, how it should be conducted, and whether it should be published or funded. The best way to promote the progress of science, on this view, is to free scientists from excessive government regulation and red tape and allow the peer review process to govern scientific funding decisions and research initiatives.

V. Bush's views have had a lasting impact on science policy. Most of the rationales developed by V. Bush play a prominent role in current debates about science funding. His idea, sometimes known as the linear model of science, continues to have a strong influence over government decisions concerning science.[1] At different times, politicians have focused on different types of practical applications of science. For example, during the Cold War, the strategic and military applications of science took center stage in the debate about science funding. After the Cold War, politicians focused more on the importance of funding science to promote economic growth and development and compete in the global economy (Guston 2000).

While most science policy analysts, scientists, and scholars accept the view that the government should invest heavily in basic research, there have been, and still are, strong arguments for a substantial government commitment to applied research, especially in the biomedical sciences. One reason that the government should invest in applied research in biomedicine is to address pressing public health and safety needs that are not likely to be addressed adequately by private industry. The pursuit of profit is industry's main reason for investing in R&D. If an area of research is not likely to be profitable, then a company may not invest in it. For example, in medicine there are a number of diseases that are so rare that the market for treatments is not large enough to guarantee a reasonable return on an R&D investment. Genetic diseases like Canavan's Disease, Lesch-Nyhan Syndrome, and Adenosine deaminase deficiency qualify as orphan diseases because they affect fewer than 200,000 U.S. citizens each year. Although the Orphan Drug Act (ODA), adopted in 1983, provides private companies with incentives to develop drugs for rare diseases, these incentives are often not strong enough to encourage companies to invest R&D funds (Iribarne 2003). Other diseases may be common but still unprofitable, due to weak consumer demand. For example, many of the diseases that are common in developing nations, such as malaria, yellow fever, and tuberculosis, could be considered orphan diseases even though they do not qualify as orphan diseases under the ODA.[2] If private companies will not sponsor research on orphan diseases, then the government should pick up the slack (Resnik 2004a).

Environmental science is another area of applied research where private companies may have little incentive to invest in R&D. Companies have very little to gain (and possibly much to lose) from demonstrating that their products or activities may threaten human health, endangered species, or the environment. Although various laws in the United States and other countries require private firms to conduct research on how their products or activities affect human health, endangered species, or the environment, most companies conduct no more research than the law requires them to conduct. For example, a pesticide company will have an incentive to conduct research to obtain approval from the Environmental Protection Agency (EPA) to market its product, but it will not have an incentive to sponsor studies on the long-term effects of its product. In the United States, government agencies, such as the EPA and the NIH, sponsor environmental research that companies are not interested in sponsoring.

As noted earlier in the book, there are good reasons for the government to invest in research on the safety, efficacy, and cost-effectiveness of drugs, medical devices, and biologics (Goozner 2004; Angell 2004). Since private companies already must sponsor research and submit data to the FDA when they want to obtain approval for a new product, one might ask whether investing government money on the same type of research would be wasteful or even redundant. Why spend public money studying a drug when a private company is already studying it? The reason that government-sponsored research would not be wasteful is that private research may be biased. I have noted several times in this book that there is a very strong positive correlation between the source of funding and research results. I have also discussed some of the strategies that private companies may employ to skew the research record, such as suppression of research, nondisclosure of data, misinterpretation of data, overestimation of clinical significance, and so on. Independent, government-sponsored studies help to compensate for these potential biases and may provide researchers and clinicians with data concerning side effects, long term risks, and comparative costs, and with additional information that can be very useful in assessing a medical product.

A good example of the value of government-sponsored, clinical research is the Antihypertensive and Lipid Lowering Treatment to Prevent Heart Attack Trial, known as ALLHAT. This $80-million study, sponsored by the NIH with support from Pfizer, compared four different types of drugs to determine which one is the most effective at lowering blood pressure. The drugs in the study included an angiotensin-converting enzyme (ACE) inhibitor, sold by Merck; a calcium channel blocker, sold by Pfizer; an alpha-adrenergic blocker, also sold by Pfizer; and a generic anti-diuretic drug (or "water pill"). The study found, to the surprise of Pfizer and many investigators, that the old anti-diuretic drug was almost as good as the three other new drugs as lowering blood pressure in many patients, and that the old drug was actually better at preventing heart disease and stroke. The study also showed that patients could save billions of dollars a year by taking generic diuretics for hypertension, since these drugs cost much less than the new, patented drugs (Angell 2004). The study yielded results that were not favorable to the pharmaceutical companies that sell the new drugs. Indeed, if a private company had sponsored this study, instead of the NIH, it is conceivable that the company would have tried to suppress, misrepresent, or "spin" these results.

8.2 Peer Review

Most of the research sponsored by the government is peer reviewed. Government agencies that sponsor research, such as the NIH, the NSF, or the DOE, have peer review committees or panels that make recommendations concerning the funding of research proposals. For many years, scientists and the public have held peer review in high regard. Peer review, according to a common view, is what allows science to be objective, rational, self-correcting, and progressive (Kitcher 1993). Empirical research conducted in the last two decades has destroyed this popular myth about the objectivity, reliability, and fairness of peer review. There is now considerable evidence that peer review can be biased, unreliable, and unfair (Shamoo and Resnik 2003; Chubin and Hackett 1990). Reviewers and editors often make irresponsible comments and judgments and may also fail to catch obvious mistakes. Peer review may not have much of an impact on the quality of research. Despite these shortcomings, most scientists and scholars agree that peer review is far superior to any other method of evaluating scientific research that one might imagine. As we shall see below, attempts to bypass the peer review system can lead to results that are far more biased than results produced by peer review. It may not be perfect, but peer review is the process that's available. Instead of using these documented problems with peer review as a reason to abandon the system, we should use them to motivate positive changes in the peer review system.

To understand how ethical problems can arise in peer review of research-funding applications, it will be useful to describe the peer review processes used by most federal funding agencies. Instead of discussing all of the different peer review procedures at different funding agencies, this section will discuss funding decisions at the NIH, which are similar to decisions made at other federal agencies. The NIH uses peer review panels, known as study sections, that make recommendations to the NIH Advisory Council. Although the advisory council usually follows the recommendations of the study sections, it has the final authority to approve or not approve research proposals. The NIH study sections usually consist of eighteen to twenty members who are experts in a particular field of research. The NIH often invites people who have had research funded to serve on study sections (Shamoo and Resnik 2003). The NIH usually assigns one primary reviewer for each research proposal and one or two secondary reviewers. The primary reviewer's job is to summarize the proposal and present it to the rest of the members of the study section. He or she should also criticize the proposal and make a recommendation concerning its overall rating. Other members of the study section also read the proposal, comment on it, and rate it. Although every member of the study section has an equal vote on any proposal, the primary and secondary reviewers can have a great deal of influence over the voting. Reviewers usually consider the following factors when evaluating research proposals (Shamoo and Resnik 2003):

1. The proposal's scientific significance
2. The proposal's methodology
3. The qualifications of the principal investigators and other scientists
4. The data or prior research that supports the proposal
5. Institutional support for the proposal and available resources
6. The proposal's budget
7. The proposal's compliance with federal laws and regulations
8. The social impact of the proposal

U.S. citizens can have some input into the NIH's decisions through their elected representatives. The NIH, like any other part of the government, is publicly accountable (Resnik 2001a). The NIH is part of the executive branch of the government. The president has oversight authority over the NIH. The president can issue executive orders that control or limit the actions of the NIH, including funding decisions. As noted in chapter 1, in 2001 President Bush decided that the NIH would fund embryonic stem cell research only on cell lines that had already been developed from embryos left over from attempted in vitro fertilization (Bruni 2001). President Bush approved about $25 million dollars a year for this research. Prior to Bush's decision, President Clinton had authorized the use of NIH funds to study but not derive human embryonic stem cells.

Congress also has oversight authority over the NIH because Congress approves the NIH's budget. Congress can also pass legislation that regulates research sponsored by the NIH. For example, in 1997, Congress passed a law forbidding the use of NIH funds to create embryos or fetuses for research purposes (Green 2001). Congress has also adopted laws that govern human experimentation, which serve as basis for federal research regulations (Resnik 2004d).

It is worth noting that individual states can make their own funding decisions concerning R&D. Although states do not have as much money as the federal government, they have often invested money in R&D initiatives designed to promote commerce and industry. As noted in chapter 1, in 2004, California citizens approved a proposition to spend $3 billion on embryonic–stem cell research over a ten-year period. Defenders of the amendment argued that it would promote important research not sponsored by the federal government and that it would help the state's economy by attracting biotechnology companies (D. Murphy 2004). In North Carolina, the state legislature has spent several million dollars a year since 1981 to support the North Carolina Biotechnology Center, a private, non-profit corporation that provides funds to new biotechnology companies and that sponsors biotechnology research and education (North Carolina Biotechnology Center 2004). Since the 1980s, many other states have made similar investments in biotechnology initiatives as well as in programs designed to promote other industries, such as information technology and pharmaceuticals.

There are also several ways that U.S. citizens can influence the NIH's decisions by communicating directly with the agency. First, patient advocacy groups frequently lobby Congress as well as the NIH concerning funding priorities. For example, since the late 1980s, HIV/AIDS activists have been very successful at securing funding for research into the diagnosis, etiology, pathology, prevention, and treatment of HIV/AIDS. The U.S. government now spends more money on research on HIV/AIDS than on any other disease. In the 1980s, women's-health groups also successfully lobbied Congress and the NIH for more research on women's health and for inclusion of women in clinical trials (Dresser 2001). Second, the directors of the NIH's various institutes, as well as the NIH director, solicit advice on funding priorities from patient advocacy groups, concerned citizens, professional associations, business leaders, and other federal agencies. Third, the NIH has established the Council of Public Representatives (COPR), composed of laypeople, which advises the NIH director on funding priorities. Fourth, the NIH also includes some laypeople on study sections (Resnik 2001a).

Even though politicians and the public have considerable input into the NIH's funding priorities, they normally do not have much influence over the peer review process. In order to make fair, effective, and objective decisions, the public delegates decision-making authority to the NIH and its administrators and reviewers. The whole point of peer review of research proposals is to have a system of decision making based on scientific, rather than political, criteria (Resnik 2001a). As demonstrated by the congressional examination of 198 NIH grants on transmitting HIV through sex or intravenous drug use, discussed in chapter 1, it is not always possible to protect the peer review process from political manipulation. Even though granting agencies tend to make decisions based on scientific criteria, the peer review system established by the NIH (and other agencies) leaves some room for considering social significance in evaluating grant proposal (Shamoo and Resnik 2003). However, considering the social significance of a proposal as a factor in making a funding decision is very different from using social significance as the overarching factor. Granting agencies should not subject research proposals to social or political litmus tests.

While scientists, politicians, and laypeople all have some input into scientific-funding decisions by the government, the peer review system used in federal agencies requires the appropriate balance of scientific decision making and public oversight. Ethical problems can arise when the public has too much or too little input. Chapter 2 argued that scientists have social responsibilities, and that these duties entail an obligation to conduct socially relevant research. If scientists do not receive adequate public input, they might not pay enough attention to an important social problem, such as HIV/AIDS or women's health. To meet their social responsibilities, scientists must seek and heed the public's advice concerning funding priorities and the selection of research problems. However, there is also the danger that powerful, well-organized advocacy groups will have a disproportionately large impact on science funding. Rebecca Dresser, a Washington University Law

Professor and member of the President's Council on Bioethics asks, rhetorically:

> As advocates lobby for funding and policy actions, will biomedical research be increasingly governed by considerations similar to those that govern highway construction and the dairy industry? Will research be diverted from the general public health needs and toward the health problems of the wealthy and powerful? If advocacy becomes a major factor in research decision making, can the process be designed to ensure that the interests of disadvantaged groups and individuals are represented? (2001, 7)

We should heed Dresser's concern about the effects of interest-group politics on the allocation government research funds. Too much input from powerful advocacy groups can lead to unfair outcomes. To avoid this problem, scientists need the appropriate amount of input from the public, and they also need to give special consideration to the needs of people who are not effective advocates for their needs, such as children, mentally ill people, and impoverished people (Resnik 2001a). Ethical problems can also arise when the public exceeds its advisory authority and interferes with the peer review process. When social and political factors trump scientific ones, peer review can be undermined.

Few people would question the public's right to oversee government funding of scientific research. The public should be able to determine, in a general way, how, where, and why its money will be spent. However, attempts to micromanage a funding agency's peer review mechanisms can cause serious harm to government-funded research. As noted in chapter 1, religious conservatives tried to stop NIH grants pertaining to research on behaviors that increase the transmission of HIV. There are two reasons to try to avoid micromanagement of the peer review process by the public. First, it can undermine the progress of science. Research on sexual behavior has always been controversial. After he was unable to secure public funds, Alfred Kinsey funded his groundbreaking research on human sexuality through a private charity that he established. If Kinsey had not been able to secure private funding, research on human sexuality would have been set back many years. Second, micromanagement of peer review can undermine the objectivity of research, since a decision to block the funding of a project is a decision to remove a decision from the scientific arena and place it in the public forum. When this happens, factors that are designed to promote objectivity and progress in research, such as empirical support, precision, and originality, become irrelevant (Leshner 2003).

Third, micromanagement of peer review by the public can undermine scientific openness. Scientists who are worried about political attacks on their work may be less likely to share their data, ideas, or results with people outside their research group, out of fear of harassment or intimidation. A researcher might decide that it is better to not inform the public about his research than to risk irritating a political-interest group that could jeopardize his funding. In 1999, Congress passed an amendment to the Freedom of Information Act that requires researchers who are receiving public funds to provide the public with access to their data. The requirement only covers published data and not preliminary data or confidential data pertaining to medical research (Kaiser 1999). Even so, many researchers have been concerned that corporations or interest groups could use the amendment to harass or intimidate researchers. For example, an animal-rights group could gain access to data on experiments involving animals in order to threaten or harass the investigators, their institution, or the NIH. An industry group could gain access to data on the effects of pollutants on humans and the environment in order to develop an attack on the research (MacIlwain 1999).

8.3 Earmarked Funds

As noted in chapter 1, the U.S. government also sponsors a great deal of scientific research via funds earmarked for special projects or special programs. Although scientists may have some input, the funding decisions are usually the product of political deal-making rather than peer review. Most earmarked projects not only bypass peer review, they also elude congressional examination because they are usually last-minute additions to omnibus spending bills (Greenberg 2001). Scientific pork barrel is politically similar to other types of pork barrel that Congress doles out each year. The political purpose of pork barrel is to allocate federal funding to a particular part of the country. Members of Congress engage in pork barrel politics to bring home the bacon to their constituencies. It makes little difference to the politicians whether the money supports a new bridge or a new laboratory, since both create jobs and impact the local economy.

Pork barrel science funding, like public interference in the peer review process, can undermine scientific objectivity because decisions about what counts as "good" science may be made by people who know little about science. Politicians may decide to approve a project that would not make it through peer review. A good example of this is the establishment of the Office of Complementary and Alternative Medicine in 1992. Responding to public pressure for more funding on complementary and alternative medicine (CAM), Congress required the NIH to establish the Office of CAM, which is now funded at $20 million a year. Many scientists protested the establishment of the Office of CAM as a waste of resources. Because millions of people use CAM remedies, there are legitimate scientific and social reasons to study the safety and efficacy of these treatments (Institute of Medicine 2005). Even so, it is not likely that NIH would have devoted an entire office to CAM if the proposal had gone through normal peer review channels.

Another example of pork barrel science is $10 million allocated for European AIDS-vaccine research, which the U.S. Congress included in last-minute budget negotiations (J. Cohen 2003). Republican members of Congress wanted to earmark this money because they wanted to increase cooperation between Italy and the United States concerning AIDS research, reward Italy for supporting the war in Iraq, and satisfy a Republican donor who now serves as an ambassador to Italy. Democrats protested that this money would never be approved through normal channels. Many scientists, including AIDS researcher Robert Gallo, questioned the science behind the vaccine research. Gallo was shocked to learn about the $10 million allocation and thought that the money would be better spent on U.S. AIDS-vaccine programs. "How do you think we feel," asked Gallo, "when we can't get support, but they can come over, do some politics, and get funding?" (J. Cohen 2003, 1640).

Even though funding via peer review is preferable to pork barrel funding, there are some good reasons to occasionally earmark funds for specific projects. First, a scientific project may be too large, costly, or interdisciplinary to fit within the normal confines of peer review. Some notable scientific projects, such as the Human Genome Project, the Space Shuttle, and the Hubble Telescope, have not fit within the normal confines of peer review. Second, if a scientific project raises national security issues, it may be unwise to expose the project to the openness and publicity that one finds in peer review (or congressional debate). For example, the Manhattan Project was approved via earmarking. Public debate of this project would have allowed the Germans and Japanese to discover the extent of the United States' atomic weapons program. Many scientific projects with military or strategic significance, such as the Strategic Defense Initiative, do not undergo normal peer review. Indeed, the U.S. government has a secret budget for national security and defense-related research that is not listed in the federal budget. Although there is no official estimate of the "black budget," some claim that it may be as high as $30 billion a year (Patton 1995).[3]

While it is often appropriate to bypass the normal peer review channels in the allocation of government research funds, this does mean that one should do without any scientific input or advice. Scientific input and advice should still be sought for projects that are being considered to receive earmarked funds. Ideally, legislators who intend to fund a science project through earmarking should submit the project to an independent scientific board that can evaluate the project and assign it a ranking in relation to other projects. Political debate about the project should include comments from the committee and its ranking of the project. While the legislators would be free to accept or reject the committee's recommendations, they would at least have the benefit of scientific expertise and advice.

8.4 Chasing Federal Dollars

Legislators are not the only people chasing federal dollars; university administrators and researchers also pursue contracts and grants with obsession and vigor. Virtually every academic institution aspires to increase the amount of revenue from government contracts and grants (Bok 2003). After World War II, when the U.S. government began to increase its funding of academic research, universities began to rely more heavily on federal funds. According to Clark Kerr, the author of the *The Uses of the University* and first president of the University of California system, "American universities have been changed almost as much by the federal research grant as by the land idea" (2001, 37).[4]

The Carnegie Classification System, which most high-level university and college administrators and trustees use in their strategic and long-term planning, has a variety of classifications, including Research University I, Research University II, Doctoral University I, Doctoral University II, Master's College/University I, and Master's College/University II. The amount of federal grant dollars received plays a key role in the Carnegie ratings. Carnegie Research I Universities receive $40 million or more in federal research support, Carnegie Research II Universities receive $15.5 to $40 million in federal research support, and other classifications receive less federal research support. In 1994, there were thirty-seven Carnegie I Research University and eighty-nine Carnegie II Research Universities (Carnegie Foundation 2000). Presidents, vice presidents, deans, and other high-level administrators, who are working for institutions that are not at the top of the Carnegie classifications, aspire to procure more federal money to push their schools to the next level. Trustees have similar ambitions.

Interference with teaching, advising and mentoring. The obsessive pursuit of government contracts and grants, like the quest for private money, can have adverse consequences for academic science. The first problem is that the emphasis on generating government dollars can undermine teaching, advising, and mentoring (Bok 2003). University hiring, promotion, and tenure decisions for scientists are usually based the quantity of publications and the amount of research support. To succeed in academic science, researchers must spend most of their time designing research, managing the research team or laboratory, writing research proposals, drafting manuscripts or presentations, and managing grants or contracts. Researchers also spend some time corresponding with colleagues and attending to administrative duties, such as serving on committees. This leaves very little time, if any, for teaching, mentoring, and supervising students. In most universities, graduate students teach introductory classes and also many intermediate ones. An undergraduate student may have very little contact with an established professor until his or her junior or senior year. Graduate students may receive poor mentoring or supervision when senior scientists are spending most of their time on sponsored research or other duties. There is mounting evidence that poor mentoring, advising, and supervision are risk factors for scientific misconduct and other ethical problems in science (National Academy of Sciences 1997).

Violations of ethical norms. A second problem with the obsessive pursuit of contracts and grants is that the emphasis on procuring government money can also tempt some researchers to violate ethical norms, such as prohibitions against data fabrication or falsification, in order to obtain funding, to continue to receive funding, or to produce tangible results. For example, Eric Poehlman, a researcher who had been working for the College of Medicine at the University of Vermont, admitted to falsifying data on fifteen federal grant applications (Kintisch 2005). Poehlman, who could receive up to five years in jail and a $250,000 fine, was barred for life from receiving federal research funding. Senior researchers, junior researchers, graduate students, and technicians face intense pressure to produce results in academia. Although most science scholars focus on the pressure to publish in science, there is an equally intense pressure to bring in money to the institution. If a senior scientist does not obtain funding, then he or she will usually not be able to conduct research, since most universities require researchers to obtain external funding for their research.

The pressure to obtain funding is most intense for researchers who are hired on soft money and must live from grant to grant or contract to contract (Barinaga 2000). A soft money position is an untenured job that is funded through an external contract or grant as opposed to funded directly through the university. Soft money researchers have very little job security; those who do not produce results may find themselves out of a job (Cardelli 1994). Soft money researchers include post-doctoral students as well as professors or research scientists. Some researchers have 100% of their salaries funded by soft money; others have a smaller percentage of their salaries funded by soft money. For example, faculty at Emory University's School of Public Health are required to pay for 70% of their salaries through contracts or grants (Academic Exchange 1999). Since soft money researchers usually have low salaries and no job benefits, universities can use soft money positions to save on labor costs. A university can create a soft money position to bring in grant or contract dollars and then terminate the position when the grant or contract expires. During the 1999–2000 academic year, 46% of the faculty positions at the University of California (all schools) were soft money positions, while 97% of the scholars at the University of Michigan's Institute for Social Research were on soft money (Barinaga 2000).

Financial misconduct. A third problem is that the unchecked pursuit of federal dollars by university officials can lead to financial misconduct. Financial misconduct can include embezzlement, fraud, or the mismanagement of funds (Shamoo and Resnik 2003). For example, a $200,000 grant or contract will usually be divided into different budget categories, such as salary support, materials and equipment, travel, phone/mail, and so on. If a researcher (or university official) attempts to use money from the grant or contract to pay for an item that is not covered by the grant or contract, or pays for something from one budget category that is not allocated to that category, this would be financial mismanagement. Two examples of mismanagement of funds would be paying a post-doctoral student's salary from a grant that he is not working on and using equipment money to remodel an office. Other examples of financial misconduct could include lying about the how much time or effort one is spending on a project, using a government budget to purchase personal items, and double-dipping (getting paid twice for the same job).

Fights over indirect costs. A fourth problem related to the pursuit of federal money concerns the bitter academic battles over money. Some of the most divisive fights involve the disposition of administrative/overhead expenses, known as indirect costs. Most grants or contracts pay for direct and indirect costs. Universities negotiate different indirect cost rates, also known as facility and administrative rates, with granting agencies (Marshall 1999b). For example, if a university applies for a grant with $100,000 in direct costs, and it has negotiated an indirect cost rate of 35%, then the total costs for the grant would be $135,000. Indirect costs, therefore, represent a large percentage of the money that the federal government spends in support of R&D as well as a significance source of revenue for research institutions.

University administrators regard indirect costs as a source of income for the institution. Most universities place the money from indirect costs in a general fund that they can use to support research at the institution or to pay for any other pressing need, such as faculty salaries, new administrative positions, library resources, new construction, and so on. Indeed, an important reason that university administrators and trustees are so keen on obtaining government grants or contracts is they know that they can use this money to support the institution. However, investigators and department heads also want to grab some of this "slush fund" to pay for items that they may want or need, such as travel expenses, books, subscriptions, student stipends, and so on. Thus, investigators, department heads, and administrators frequently have fights over indirect costs.

Academic prima donnas. A fifth problem related to the desire for federal money concerns the recruitment of superstar researchers who can pull in federal dollars. Chapter 1 mentioned several cases in which universities have competed for star researchers with promises of high salaries, new academic centers, and other perks. Although there is nothing inherently wrong in offering a researcher high financial rewards, bidding wars can lead to irresponsible spending, intellectual property disputes, and salary inequities, which can undermine the institution's financial stability and morale. The competition for academic superstars is no different, in principle, from competitions among sports teams for professional athletes or competitions among universities for athletic coaches. Although most people are accustomed to competition and greed in the world of sports, it still seems out of place in the academy, given its traditional commitment to the pursuit of truth and knowledge.

To avoid these problems, university administrators, trustees, and investigators must find a way to reign in or control the obsession with federal dollars. Generating revenue from external, government sources is an important goal in research, but it is not the most important goal. Educating students, developing new knowledge and contributing to society are the most important goals for academic institutions. Money, whether from public or private sources, should be viewed as a means to those goals, not as an end in itself. Most readers would probably agree with these statements about academic priorities. Putting them into practice in universities and colleges is a different matter, however. Exploring this problem in depth would take us way beyond the scope of this book, but we can discuss paths toward developing a solution.

It is a fact of human nature that people will modify their behavior in order to receive rewards or avoid punishments. The way to control the unbridled pursuit of federal dollars in academia is to give administrators, researchers, and others rewards that are not based on the amount of money that one generates for the institution. Universities should not make the amount of revenue generated by a researcher the primary measure of that researcher's performance. Universities should give more weight to noneconomic criteria, such as the quality of published work, teaching, advising, and community service, when making hiring and promotion decisions. Likewise, universities should not make the amount of revenue generated by an administrator the primary measure of that administrator's performance. They should give more weight to noneconomic criteria, such as leadership and management skills, fairness, accessibility, accountability, and so on.

Many university officials probably already try to follow this advice, and few would admit to giving significant weight to financial criteria when making personnel and other administrative decisions. Clearly, some universities are better than others when it comes to controlling the obsession with federal money. However, it is hard to deny that somehow the pursuit of federal dollars has gotten out of hand in the last few decades, and that something needs to be done.

8.5 An Objection and Reply

One might object that all the problems with the unbridled pursuit of government money can also occur in the unbridled pursuit of private money. Moreover, privately funded research introduces many different problems, such as ownership of stock or leadership of a company, that do not arise in publicly funded research. Thus, the arguments in this chapter do not show that the pursuit of federal contracts and grants has a uniquely corrupting influence on academic science. Private money is a much worse influence on science.

There is no need to deny this point. The point here is not that government funding causes more corruption than private funding; the point is that government funding can also lead to ethical problems in science. What matters most, in research, is how scientists pursue and use money, not the source of their money.[5] It is possible to exhibit a high degree of moral integrity when conducting industry-sponsored research, or to exhibit a low degree of moral integrity when conducting government-sponsored research. Even if private-interest science tends to create more ethical problems than public-interest science, recognizing that the influence of money on public-interest science can also lead to ethical problems is an important point to make in the debate about the relationship between science and money. Many other authors have expressed grave concerns about the privatization of science, and this book echoes their distress. But it would be a big mistake to forget that public science also has ethical problems related to the acquisition, allocation, management, and expenditure of money.

NINE

CONCLUSION

Valuing Truth and Integrity in Research

> Money never made a man happy yet, nor will it. There is nothing in its nature to produce happiness. The more a man has, the more he wants. Instead of its filling a vacuum, it makes one.
>
> —Ben Franklin

This book has explored the relationship between money and the norms of science. The book will not be the last word on this complex and important topic, but hopefully it will shed some light on this ongoing debate. The first part of the book, chapters 1 to 4, explored the foundations of relationship between money and scientific norms. Chapter 1 provided some background information on economic influences and financial interest related to scientific research. It described the sources of funding for science, the interests and expectations of research sponsors and scientists, and some current practices that interfere with the norms of science. The first chapter also gave an account of some recent cases that illustrate the impact of money on science and the failure of companies to publicize the dangers of prescribing anti-depressants to children. Chapter 2 developed framework for thinking about the norms of science. It argued that science has its own norms (or values), which consist of epistemological norms and ethical norms. Epistemological norms—such as empirical adequacy, simplicity, generality, objectivity, and precision—guide scientific testing, explanation, and theory construction. Ethical norms—such as honesty, carefulness, openness, freedom, social responsibility, respect for colleagues, and respect for research subjects—guide scientific conduct.

Chapter 3 analyzed science's most important norm—objectivity—in more depth. It argued that scientists have an obligation to strive for objectivity in research because pluralistic and democratic societies, such as the United States, need to rely on beliefs and methods that are independent of political, social, cultural, and economic values. Science can provide opposing parties in public policy debates with neutral and impartial statements about the natural world or "facts." The last forty years of scholarship in the history, sociology, psychology, and philosophy of science has demonstrated that political, social, cultural, moral, and economic factors often influence scientific judgment and decision making. Objectivity in science is an ideal, not a reality. Even though scientists often fall short of this ideal, they should still strive to be objective. Chapter 4 explored the various ways that money can affect scientific norms, such as honesty, objectivity and openness. It examined the impact of financial interests and economic influences on the selection of problems, experimental design, data analysis and interpretation, the recruit of subjects, publication, and data sharing.

The second part of the book, chapters 5 to 8, examined some practical problems resulting from the impact of money on science. These chapters also proposed some policy solutions to these problems. Chapter 5 examined conflicts of interest in research, for individuals as well as for institutions. The chapter distinguished between conflicts of interest, conflicts of commitment, and conflicts of obligation. It also discussed strategies for dealing with conflicts of interest, such as disclosure, prohibition, and conflict management. Chapter 6 described the intellectual property regime found in the United States and other industrialized nations. The chapter argued that intellectual property rights are important for promoting innovation and investment in research and development, and that society must carefully balance private and public interests regarding the control intellectual property. The chapter also argued that government agencies and the courts should takes steps to prevent private corporations from abusing the intellectual property system. Chapter 7 discussed some ethical problems and issues with scientific publication that can result from financial interests in research, such as bias, undeserved authorship, duplicate publication, salami science, and problems with access to data. Chapter 8 described the U.S. government's system for funding scientific research. It described the role of peer review and public oversight in research funding, and argued that public funding of research requires an optimal balance of public and scientific input. The chapter also cautioned against denial of government funds as a means of censoring controversial research, and decried excessive use of earmarked money in research.

By now it should be clear to the reader that the main theme of this book is that money can have an adverse effect on science's adherence to epistemological and ethical norms. The thesis that money can undermine scientific norms may sound preposterous to some people and painfully obvious to others. Those who find this idea absurd probably subscribe to the popular mythology that scientists are objective and unbiased and would never allow greed or ambition to interfere with the search for the truth. Those who find the thesis painfully obvious view scientists as human beings, who, like the rest of us, can succumb to biases and temptations. The truth is probably somewhere in between. Scientists are human beings who have a strong desire to know the truth and to understand nature. Although scientists strive for objectivity, they often fall short of the mark. Money can hinder the search for truth, but it cannot stop it. As long as human beings exist, there will be people who want to strive for knowledge, truth, and understanding. These people, whom we call scientists, will continue on this path even if detours, obstacles, and temptations come their way.

Many scholars and critics who write about science and money are very good at pointing out problems but not very good at thinking about solutions. It is very easy to say, "Look at how science had become corrupted by private industry. . . . Look at how academic institutions resemble businesses. . . . Look at how many scientists have conflicts of interests. . . . Look at how many scientists are starting their own companies. . . . Look at how intellectual property rights are interfering with the sharing of information. . . . Look at how companies suppress publication," and so on, and so on, and so on.

Yes, there are problems. Indeed, there are many problems. But what are we going to do about them? Society can consider two extreme responses: eliminate private money from science (a socialistic proposal) or take a laissez-faire approach (a capitalistic proposal). The best solution, I believe, lies somewhere between these two extremes.

Eliminating private money from science is not a realistic option, since private businesses now fund more than half the world's R&D, and governments do not have enough money to support R&D at its current levels. Moreover, one might argue that private companies have a moral and legal right to sponsor R&D, since they conduct R&D to develop goods and services and market their products. If companies have a right to conduct business and advertise their business, then they have a right to conduct research that is related to their business and to publish that research. Eliminating private money from academic science is also not a realistic option, since universities now obtain a large percentage of their R&D funds from the private sector, and there are many benefits of university-industry collaborations. Additionally, as we have seen, government sponsorship of research does not remove all ethical problems related to financial interests and economic pressures; it only creates a different set of problems.

The other extreme, taking a laissez-faire approach to the relationship science and money, is not a wise idea either, because, as we have seen in this book, financial interests and economic influences can have a corrupting effect on the norms of science. The love of money can undermine the integrity of the research process and the search for truth.

The best option is to try to manage the relationship between scientific research and financial interests, and to develop social and economic institutions that support R&D and promote the values of science, such as objectivity, honesty, and openness. Universities, private companies, granting agencies, journals, and professional associations should develop rules and guidelines for mitigating money's corrupting influence on science. Some of those mentioned in this book include the following:

- Disclose and manage individual and institutional conflicts of interest and prohibit those conflicts that pose a significant threat to science or the public
- Achieve proper balance of public and private interests in intellectual property law and policy; prevent abuse of the system
- Create legislation to clarify and strengthen the research exemption in patent law
- Develop institutional rules for authorship and publication that promote responsible authorship and prevent wasteful or duplicative publication
- Develop policies for access to databases that reward researchers for their efforts but also allow the public to have access to data
- Construct public databases for unpublished data
- Mandate the registration of all clinical trials and the public release of all clinical trial data
- Experiment with different ways of financing scientific publication, such as open-access science
- Carefully review confidential disclosure agreements to avoid unreasonable delays in publication, suppression of publication, or intimidation of researchers
- Ensure that government contracts or grants for research do not bypass peer review

- Cut back on scientific earmarking and ensure that earmarked funds receive some type of scientific review
- Strive for the appropriate balance of scientific decision making and public oversight in the funding of science; avoid the micromanagement of contracts and grants by the public
- Use some government funds to conduct research on new biomedical products developed by industry in order to counteract biases present in industry-sponsored research; support clinical trials that compare two therapies
- Reign in the academy's obsession with generating revenue from government contracts and grants; emphasize nonfinancial measures of performance for researchers and administrators

And last, but not least:

- Support education, training, and mentoring on research ethics for scientists and students in all sectors of the research economy—including colleges and universities, government and private laboratories, and contract research organizations (National Academy of Sciences 1997, 2002)—which can take place through academic courses, online training modules, seminars, informal discussions, and role modeling.

Clearly, much more work needs to be done. Scientists, as well as leaders from academia, industry, and government, need to develop and refine policies and guidelines dealing with science and money. Academic institutions and government agencies should provide financial support for policy development and educational and training initiatives dealing with research ethics. Leaders in research-intensive industries, such as pharmaceuticals, biotechnology, and electronics, should publicly affirm their commitment to research ethics and should support educational, training, and mentoring activities involving ethics. Those who study science, such as historians, sociologists, anthropologists, economists, psychologists, and philosophers, need to conduct more empirical and conceptual research on the relationship between science and money. Finally, all parties who care about promoting scientific research should come to a fuller understanding of how the public views the relationship between science and money, and how one can maintain the public's trust in science. The public understands the adage "The love of money is the root of all evil." Scientists must convince the public that they love truth and integrity more than money.

NOTES

Chapter Two

1. Many readers associate the phrase "the norms of science" with the pioneering work of sociologist Robert Merton (1938, 1973). Merton developed an account of scientific norms based on years of interviews with scientists about their work. According to Merton, there are four basic norms: (1) disinterestedness, (2) universalism, (3) communalism, and (4) organized skepticism. Disinterestedness is the idea that the only interest that scientists should have is an interest in the truth; they should not conduct research to advance their personal interests or political ideologies. Universalism is the idea that the validity of one's scientific claims should not depend on one's nationality, race, gender or culture, for scientific truth is not relative to a particular culture, race, gender, or society. Communalism is the idea that the fruits of scientific investigation belong to no single person and should be jointly owned. Organized skepticism is the idea that one should critically examine all scientific beliefs and assumptions, including one's own. While sociologists, historians, and philosophers of science owe a great debt to Merton and his insights, I prefer my own account of the norms of science, which is based, in part, on some of Merton's thought. I think the norms I defend in this chapter embody most of Merton's basic ideas but also expand them; the norms I propose are more complex and subtle than Merton's norms. For example, I think of organized skepticism as actually being composed of several different norms, including honesty, openness, carefulness, consistency, testability, and rigor. I also include some norms that are not on Merton's list, such as social responsibility and respect for research subjects. I should mention that I have had some contact

with Merton about the norms of science. He has read my work on this subject and approved of my program of "consolidating ethics with the institutional sociology of science and social epistemology" (Merton 2000).

2. For further discussion that challenges this widely held view, see Resnik (1992a).

Chapter Three

1. Science can also exist in nondemocratic societies, and it may serve an important function in those societies as well. However, in some societies, science may function as an instrument of the state. Consider, for example, whether science was regarded as unbiased in the Soviet Union during Stalin's regime or in Germany during Hitler's rule. In these societies, science would play an important role in carrying out governmental policies but not an important role in helping to settle controversial issues, since these issues would not be open to public debate.

2. This book assumes what is known as the correspondence theory of truth: a statement or belief is true if and only if it accurately represents reality. True statements fit the facts. Thus, the statement "Raleigh is in North Carolina" is true if and only if the words "Raleigh" and "North Carolina" both refer to places in the world and Raleigh is in North Carolina. For further discussion, see Goldman (1986) and Resnik (1992b).

3. The relationship between "science" and "scientists" is simply this: science is a social activity or discipline conducted by scientists. Scientists can affect the characteristics of science through their actions and beliefs. For example, the physics community accepted Einstein's explanation of the photoelectric effect because physicists, as individuals, accepted his explanation.

4. For more on the observation/theory distinction as well as arguments for and against realism about scientific theories, see Klee (1997), Rosenberg (2000), Curd and Cover (1998), and Churchland and Hooker (1985).

5. There are two very different ways of reading Kant's distinction between phenomena and noumena, the "two world" interpretation and the "two aspect" interpretation. According to the "two world" view, which this chapter has implicitly accepted, phenomena and noumena are two different realms of existence. For example, a human being would contain phenomena (the body) as well as noumena (the soul). According to the "two aspect" interpretation, phenomena and noumena are two different ways of conceiving the same world. We could conceive of a human being as a body or as a soul. See Alison (2004).

Chapter Four

1. Critics of human pesticide experiments have argued that some pesticide companies underpowered their studies so that they would not be able to

detect statistically significant adverse effects on human subjects. See Lockwood (2004).

Chapter Five

1. "Judgment" is used here as a catch-all term that refers to many different cognitive functions in which a person must form an opinion about something, such as perception, communication, reasoning, and deliberation.

2. The term "misconduct," as it is used here, is much broader than the federal definition of the term, which limits the definition of misconduct to fabrication, falsification, or plagiarism. Some organizations and institutions have included serious deviations from accepted practices and serious violations of human research regulations within the definition of misconduct. See Resnik (2003b).

3. This is assuming, of course, that people know what their ethical duties are. A person could still have conflicts among different ethical duties, but this is not that same as to a have a conflict between duties and personal (or other) interests.

4. Institutional Review Boards for research on human subjects (IRBs), also known as Research Ethics Committees (RECs), have existed since the 1960s when the NIH established peer-review committees to review scientific and ethical aspects of human experimentation in its intramural research program. In 1974, the U.S. Congress passed the National Research Act, which granted federal agencies, such as the NIH and the FDA, the power to develop regulations pertaining to biomedical and behavioral research involving human subjects. The regulations adopted by these agencies established the IRB system as a procedure for overseeing human research. Research that is funded by federal agencies or privately funded research conducted to obtain FDA approval for a new drug, biologic, or medical device, must be approved by an IRB. The IRBs evaluate studies according to requirements established by federal regulations, which address such issues as informed consent, benefits and risks, safety, privacy and confidentiality, protection for vulnerable subjects, and good research design. If an IRB determines that a proposed study does not meet the legal requirements, then it should not approve the study until the researchers make changes in the study to bring it into conformity with the regulations. The IRBs are also responsible for reviewing studies that they have approved, providing education and guidance to researchers, and auditing and monitoring research. The U.S. regulations require that IRBs be composed of members with different backgrounds and expertise, including institutional members and noninstitutional members, and scientific members and nonscientific members. Many other countries have laws similar to the United States' laws pertaining to the conduct of human research, and these laws also require that research be reviewed by an IRB. International ethics guidelines, such as the

Nuremberg Code, the World Medical Association's Helsinki Declaration, and the Council for the International Organization of Medical Sciences' guidelines, also address the conduct of human subjects research (Brody 1998; Emanuel et al. 2004).

5. In addition the ethical problems with excessively high fees for enrolling patients in clinical trials, researchers also face potential legal problems related to payments for referral and fraud. In the United States, a variety of federal and state laws prohibit physicians from receiving fees (or "kickbacks") for referrals (Studdert, Mello, and Brennan 2004).

Chapter Six

1. This book will focus on U.S. patent law, which is similar to the patent law found in other industrialized nations, such as countries that belong to the European Union.

2. The upstream/downstream distinction is relative to a particular area of scientific or technical field. A technology is an upstream technology relative to other downstream technologies if it can be used in making or developing these other technologies. For example, the transistor is an upstream technology relative to the integrate circuit, but the integrated circuit is an upstream technology related to a computer chip.

3. When scientists are conducting research for private companies or government agencies or are collaborating with scientists at other institutions, they often sign contracts pertaining to intellectual and tangible property known as confidential disclosure agreements (CDAs) and material transfer agreements (MTAs). A CDA is an agreement to not disclose proprietary or confidential information. Parties to a CDA may include the university or a private company, and its agents or employees, such as scientists, graduate students, and technicians. Parties who sign such an agreement are legally bound to not disclose the confidential information and to make a reasonable effort to maintain confidentiality. Since publication is a form of disclosure, universities that sign CDAs often try to limit the period of agreement to allow scientists to publish their work. For example, a scientist conducting research on a new drug for a company might agree to not disclose his work without the company's permission for six months. A CRADA is an agreement that a government agency signs with a private company and/or university to develop and commercialize technologies discovered or invented under a government contract or grant. An MTA is an agreement pertaining to the transfer of research materials, such as chemicals, reagents, viruses, tissues, transgenic mice, and electronic devices, between different institutions or organizations. An MTA defines allowable uses of these materials and prohibits a party from transferring materials to someone else without permission of the other party. Materials may be provided at no cost or for a fee. The recipient usually must also pay for shipping costs.

4. In this book, I will focus only on the relationship between IPRs and values related to the progress of science and technology, such as openness.

I will not consider broader, moral concerns here. See Resnik (2004a,b); Andrews and Nelkin (2000); and Magnus, Caplan, and McGee (2002) for further discussion.

Chapter Seven

1. See note 3, chapter 6.

Chapter Eight

1. One might argue that the model is an oversimplification of the process of research and development. First, sometimes the time period between basic research and practical applications is very long. For example, I suspect it will take many decades or even centuries before we can develop practical applications from research on black holes or string theory. Second, sometimes practical or technical innovations lead to advances in basic research, instead of the other way around. For example, cell theory was developed following advances in microscopy. Without good microscopes, it was not possible to observe cells. Third, as I have noted several times in this book, the line between basic and applied research is often not clear, especially in fields such as biotechnology, genomics, and computer and information science.

2. According to some estimates, 90% of the world's biomedical R&D funds are spent on conditions that affect only 10% of the world's disease burden (Francisco 2002).

3. There are many political and moral problems with black budgets, such as lack of accountability and the promotion of a culture of secrecy within the government, which will not be discussed in this book. See Foerstel (1993).

4. The "land grant" idea refers to the donation of land to state-chartered universities by the states. The universities could use the land for construction, or they could sell or lease the land to raise revenues. For example, the University of Wyoming, a land-grant institution, was established in 1890 by the State of Wyoming through a gift of several thousand acres of land in Laramie, Wyoming, and the surrounding area.

5. An assumption is made here that the R&D funds do not come from a corrupt source, such as a money laundering or counterfeiting operations. There is no need to respond to the objection that the government steals its money or that private companies extort theirs.

REFERENCES

AAMC (Association of American Medical Colleges). 1990. Guidelines for dealing with conflicts of commitment and conflicts of interest in research. *Academic Medicine* 65:491.

AAU (Association of American Universities). 2001. Report on individual and institutional financial conflict of interest. http://www.aau.edu/research/COI.01.pdf (accessed May 9, 2006).

Academic Exchange. 1999. Resources, risk, and reward. September 1999. http://www.emory.edu/ACAD_EXCHANGE/1999/sept99/resources.html (accessed December 11, 2003).

ACTR (Australian Clinical Trials Registry). 2005. Frequently Asked Questions. http://www.actr.org.au/faq.aspx#1 (accessed May 9, 2006).

Allison, H. 2004. *Kant's transcendental idealism*, revised edition. New Haven, CT: Yale University Press.

Als-Nielsen, B., W. Chen, C. Gluud, and L. Kjaergard. 2003. Association of funding and conclusions in randomized drug trials. *Journal of the American Medical Association* 290:921–28.

American Chemical Society. 1995. *Will science publishing perish? The paradox of contemporary science journals.* Washington, DC: American Chemical Society.

———. 2000. Ethical guidelines to publication of chemical research. https://paragon.acs.org/paragon/ShowDocServlet?contentId=paragon/menu_content/newtothissite/eg_ethic2000.pdf (accessed May 9, 2006).

American Heritage Dictionary. 2001. http://dictionary.com (accessed May 9, 2006).

American Physical Society. 2002. Guidelines for professional conduct. http://www.aps.org/statements/02_2.cfm. (accessed May 9, 2006).

American Society for Biochemistry and Molecular Biology. 1998. Code of ethics. http://www.asbmb.org/ASBMB/site.nsf/web/035D570E3A8E81FA85256C7C00535A61?opendocument (accessed May 9, 2006).

Andrews, L., and D. Nelkin. 2000. *Body bazaar.* New York: Crown Publishers.

Angell, M. 2000. Is academic medicine for sale? *The New England Journal of Medicine* 342:1516–18.

———. 2004. *The truth about drug companies.* New York: Random House.

Association of American Medical Colleges. *See* AAMC.

Association of American Universities. *See* AAU.

Associated Press. 2005. Chronology of events surrounding Vioxx. *New York Times* (November 3, 2005): A1.

Ayer, A. 1952. *Language, truth, and logic.* New York: Dover.

Babbage, C. 1830. *Reflections on the decline of science in England.* New York: Augustus Kelley, 1970.

Bacon, F. 1626/1996. *The new Atlantis.* http://oregonstate.edu/instruct/phl302/texts/bacon/atlantis.html (accessed May 9, 2006).

Baker v. Seldon. 101 U.S. 99 (1879).

Bailar, J. 1997. The promise and problems of meta-analysis. *New England Journal of Medicine* 337:559–61.

Barinaga, M. 1999. No winners in patent shootout. *Science* 284:1752–53.

———. 2000. Soft money's hard realities. *Science* 289:2024–28.

Benatar, S. 2000. Avoiding exploitation in clinical research. *Cambridge Quarterly of Healthcare Ethics* 9:562–65.

Berkeley, G. 1710. *Principles of human knowledge.* New York: Oxford University Press, 1996.

Biotechnology Industry Organization (BIO). 2003. History of BIO. http://www.bio.org/aboutbio/history.asp (accessed May 9, 2006).

Blackburn, E. 2004. Bioethics and the political distortion of biomedical science. *New England Journal of Medicine* 350:1379–80.

Black's Law Dictionary. 1999. 7th ed. St. Paul, MN: West Publishing.

Bloor, D. 1991. *Knowledge and social imagery.* 2nd ed. Chicago: University of Chicago Press.

Blumenstyk, G. 2003. Colleges report $827 million in 2001 royalties. *Chronicle of Higher Education* 49, no. 39 (May 22, 2003): A28.

Blumenthal, D., E. Campbell, M. Anderson, N. Causino, and K. Louis. 1997. Withholding research results in academic life science: Evidence from a national survey of faculty. *Journal of the American Medical Association* 277:1224–28.

Blumenthal, D., E. Campbell, N. Causino, and K. Louis. 1996. Participation of life science faculty in research relationships with industry. *New England Journal of Medicine* 335:1734–39.

Blumenthal, D., M. Gluck, K. Louis, and D. Wise. 1986. Industrial support of university research in biotechnology. *Science* 231:242–46.

Bodenheimer, T. 2000. Uneasy alliance: Clinical investigators and the pharmaceutical industry. *The New England Journal of Medicine* 342:1539–44.

Bok, D. 2003. *Universities in the marketplace.* Princeton, NJ: Princeton University Press.

Bombardier, C., L. Laine, A. Reicin, D. Shapiro, R. Burgos-Vargas, B. Davis, R. Day, M. Ferraz, C. Hawkey, M. Hochberg, T. Kvien, T. Schnitzer, and the VIGOR Study Group. 2000. Comparison of upper gastrointestinal toxicity of rofecoxib and naproxen in patients with rheumatoid arthritis. VIGOR study group. *New England Journal of Medicine* 343:1520–28.

Bonner, L. 2003. Star draws light, new lab to ECU. *News and Observer* (September 22, 2003): A1, A10.

Booth, W. 1988. Conflict of interest eyed at Harvard. *Science* 242:1497–99.

Bowie, N. 1994. *University-business partnerships: An assessment.* Lanham, MD: Rowman and Littlefield.

Boyd, R. 1984. The current status of scientific realism. In *Scientific Realism*, ed. J. Leplin, 41–83. Berkeley: University of California Press.

Bradley, G. 2000. Managing conflicting interests. In *Scientific Integrity*, ed. F. Macrina, 131–56. 2nd ed. Washington, DC: American Society for Microbiology Press.

Brenner v. Manson. 383 U.S. 519 (1966).

Bresalier, R., R. Sandler, H. Quan, J. Bolognese, B. Oxenius, K. Horgan, C. Lines, R. Riddell, D. Morton, A. Lanas, M. Konstam, J. Baron, and the Adenomatous Polyp Prevention on Vioxx (APPROVe) Trial Investigators. 2005. Cardiovascular events associated with rofecoxib in a colorectal adenoma chemoprevention trial. *New England Journal of Medicine* 352:1092–1111.

Broad, W., and N. Wade. 1993. *Betrayers of the truth.* New York: Simon and Schuster.

Brody, B. 1998. *The ethics of biomedical research: An international perspective.* New York: Oxford University Press.

Brown, J. 2000. Privatizing the university—The new tragedy of the commons. *Science* 290:1701–2.

Bruni, F. 2001. Decision helps define the president's image. *New York Times* (August 10, 2001): A1.

Bureau of Economic Analysis. 2006. Gross domestic product. http://www.bea.gov/bea/dn/home/gdp.htm (accessed May 9, 2006).

Bush, V. 1945. *Science: The endless frontier.* Washington, DC: National Science Foundation, 1990.

Butler, D. 1999. The writing is on the web for science journals in print. *Nature* 397:195–200.

Callaham, M., R. Wears, E. Weber, C. Barton, and G. Young. 1998. Positive-outcome bias and other limitations in the outcome research abstracts submitted to a scientific meeting. *Journal of the American Medical Association* 280:254–57.

Campbell, E., B. Clarridge, M. Gokhale, L. Birenbaum, S. Hilgartner, N. Holtzman, and D. Blumenthal. 2002. Data withholding in academic genetics: Evidence from a national survey. *Journal of the American Medical Association* 287:473–80.

Cardelli, J. 1994. Confronting the issues and concerns facing non-faculty (soft-money) astronomers. http://www.astronomy.villanova.edu/faculty/ara/ara_art.htm (accessed May 9, 2006).

Carey, B. 2004. Long after Kinsey, only the brave study sex. *New York Times* (November 9, 2004): A1.

Carnap, R. 1950. Empiricism, semantics, and ontology. *Revue Internationale de Philosophie* 4:20–40.

Carnegie Foundation. 2000. The 2000 Carnegie classification: Background and description. http://www.carnegiefoundation.org/classifications/index.asp?key=785 (accessed May 12, 2006).

Celera. 2003. Our history. http://www.celera.com/celera/history (accessed May 12, 2006).

Cho, M., and L. Bero. 1996. Quality of drug studies published in symposium proceedings. *Annals of Internal Medicine* 124:485–89.

Cho, M., R. Shohara, A. Schissel, and D. Rennie. 2000. Policies on faculty conflicts of interest at U.S. universities. *Journal of the American Medical Association* 284:2203–08.

Chubin, D., and E. Hackett. 1990. *Peerless science*. Albany, NY: SUNY Press.

Churchland, P., and C. Hooker. 1985. *Images of science: Essays on realism and empiricism*. Chicago: University of Chicago Press.

Clery, D. 2004. Mixed week for open access in the U.K. *Science* 306:1115.

ClinicalTrials.gov. 2006. About ClinicalTrials.gov. http://clinicaltrials.gov/ (accessed May 9, 2006).

Cohen, J. 1994. U.S.-French patent dispute heads for showdown. *Science* 265:23–25.

———. 2003. Earmark draws criticism, creates confusion. *Science* 302:1639–40.

Cohen, S., and H. Boyer. 1980. Process for producing biologically functional molecular chimeras. U.S. Patent 4:237, 224.

Collins, F., and V. McKusick. 2001. Implications of the human genome project for medical science. *Journal of the American Medical Association* 285:540–44.

Collins, F., M. Morgan, and A. Patrinos. 2003. The human genome project: Lessons from large-scale biology. *Science* 300:286–90.

Committee on Publication Ethics. 2005. A code of conduct for editors of biomedical journals. http://www.publicationethics.org.uk/guidelines/code (accessed November 15, 2005).

Congressional Budget Office. 2003. The long-term implications of current defense plans: Summary update for fiscal year 2004. http://www.cbo.gov/showdoc.cfm?index=4449&sequence=0#pt3 (accessed May 12, 2006).

Couzin, J. 2004. Legislators propose a registry to track clinical trials from start to finish. *Science* 305:1695.

Curd, M., and J. Cover, eds. 1998. *Philosophy of science.* New York: W. W. Norton.

Daley, G. 2004. Missed opportunities in embryonic stem-cell research. *New England Journal of Medicine* 351:627–28.

Dalton, R. 2001. Peers under pressure. *Nature* 413:102–4.

Dana, J., and G. Loewenstein. 2003. A social science perspective on gifts to physicians from industry. *Journal of the American Medical Association* 290: 252–55.

Davidson, R. 1986. Source of funding and outcome of clinical trials. *Journal of General Internal Medicine* 1:155–58.

Davis, M. 1982. Conflict of interest. *Business and professional ethics journal* 1 (4): 17–27.

Davis, R., and M. Mullner. 2002. Editorial independence at medical journals owned by professional associations: A survey of editors. *Science and Engineering Ethics* 8:513–28.

Daubert v. Merrell Dow Pharmaceuticals. 509 U.S. 579 (1993).

De Angelis, C. 2000. Conflict of interest and the public trust. *Journal of the American Medical Association* 284:2237–38.

De Angelis, C., J. Drazen, F. Frizelle, C. Haug, J. Hoey, R. Horton, S. Kotzin, C. Laine, A. Marusic, A. Overbeke, T. Schroeder, H. Sox, and M. Van Der Weyden. 2004. Clinical trial registration: A statement from the international committee of medical journal editors. *New England Journal of Medicine* 351:1250–51.

Demaine, L., and A. Fellmeth. 2002. Reinventing the double helix: A novel and nonobvious reconceptualization of the biotechnology patent. *Stanford Law Review* 55:303–462.

———. 2003. Natural substances and patentable inventions. *Science* 300:1375–1376.

Derwent Information. 2001. Frequently asked questions. http://www.derwent.com/ (accessed May 12, 2006).

Dewey, J. 1910. *The influence of Darwin on philosophy and other essays in contemporary thought.* New York: Henry Holt and Company.

Diamond v. Diehr. 450 U.S. 175 (1981).

Diamond v. Chakrabarty. 447 U.S. 303 (1980).

Dickersin, K., and D. Rennie. 2003. Registering clinical trials. *Journal of the American Medical Association* 290:516–23.

Doll, J. 1998. The patenting of DNA. *Science* 280:689–90.

Drazen, J., and G. Curfman. 2002. Financial associations of authors. *New England Journal of Medicine* 346:1901–2.

Dresser, R. 2001. *When science offers salvation: Patient advocacy and research ethics.* New York: Oxford University Press.

Ducor, P. 2000. Coauthorship and coinventorship. *Science* 289:873–74.

Dupré, J. 1993. *The disorder of things: Metaphysical foundations of the disunity of science.* Cambridge: Cambridge University Press.

Easterbrook, P., J. Berlin, R. Gopalan, and D. Matthews. 1991. Publication bias in clinical research. *Lancet* 337:867–72.

Eisenberg, R. 2003. Patent swords and shields. *Science* 299:1018–19.

Emanuel, E., and F. Miller. 2001. The ethics of placebo-controlled trials—A middle ground. *New England Journal of Medicine* 345:915–19.

Emanuel, E., R. Crouch, J. Arras, J. Moreno, and C. Grady, eds. 2004. *Ethical and regulatory aspects of clinical research.* Baltimore, MD: Johns Hopkins University Press.

Environmental Protection Agency (EPA). 1994. Setting the record straight: Secondhand smoke is a preventable health risk. http://www.epa.gov/smokefree/pubs/strsfs.html (accessed November 9, 2005).

FDA (Food and Drug Administration). 1999. Financial disclosure by clinical investigators. 21 Code of Federal Regulations 54.1–54.6

———. 2002. FY 2002 PDUFA financial report. http://www.fda.gov/oc/pdufa/finreport2002/financial-fy2002.html (accessed November 17, 2003).

Feist Publications Inc. v. Rural Telephone Service. 499 U.S. 340 (1991).

Fine, A. 1996. *The shaky game: Einstein, realism, and quantum mechanics.* Chicago: University of Chicago Press.

Flanagin, A., L. Carey, P. Fontanarosa, S. Phillips, B. Pace, G. Lundberg, D. Rennie. 1998. Prevalence of articles with honorary and ghost authors in peer-reviewed medical journals. *Journal of the American Medical Association* 280:222–24.

Fletcher, R., and S. Fletcher. 1997. Evidence for the effectiveness of peer review. *Science and Engineering Ethics* 3:35–50.

Foerstel, H. 1993. *Secret science: Federal control of American science and technology.* New York: Praeger Publishers.

Food and Drug Administration. *See* FDA.

Foster, F., and R. Shook. 1993. *Patents, copyrights, and trademarks.* 2nd ed. New York: John Wiley and Sons.

Francisco, A. 2002. Drug development for neglected diseases. *The Lancet* 360:1102.

Friedberg, M., B. Saffran, T. Stinson, W. Nelson, and C. Bennett. 1999. Evaluation of conflict of interest in new drugs used in oncology. *Journal of the American Medical Association* 282:1453–57.

Gelijns, A., and S. Their. 2002. Medical innovation and institutional interdependence: Rethinking university-industry connections. *Journal of the American Medical Association* 287:72–77.

Gibbs, W. 1996. The price of silence: Does profit-minded secrecy retard scientific progress? *Scientific American* 275 (5): 15–16.

Gibbard, A. 1990. *Wise choices, apt feelings.* Cambridge, MA: Harvard University Press.

Giere, R. 1988. *Explaining science.* Chicago: University of Chicago Press.

———. 2004. *Understanding scientific reasoning*. 4th ed. Belmont, CA: Wadsworth.

Goldhammer, A. 2001. Current issues in clinical research and the development of new pharmaceuticals. *Accountability in Research* 8:283–92.

Goldman, A. 1986. *Epistemology and cognition*. Cambridge, CA: Harvard University Press.

———. 1999. *Knowledge in social world*. Oxford: Oxford University Press.

Goozner, M. 2004. *The $800 million pill*. Berkeley: University of California Press.

Grassler, F., and M. Capria. 2003. Patent pooling: Uncorking a technology transfer bottleneck and creating value in the biomedical research field. *Journal of Commercial Biotechnology* 9 (2): 111–19.

Green, R. 2001. *The human embryo research debates: Bioethics in the vortex of controversy*. New York: Oxford University Press.

Greenberg, D. 2001. *Science, money, and politics*. Chicago: University of Chicago Press.

Gregory, C., S. Morrissey, and J. Drazen. 2003. Notice of duplicate publication. *New England Journal of Medicine* 348:2254.

Guenin, L. 1996. Norms for patents concerning human and other life forms. *Theoretical Medicine* 17:279–314.

Guston, D. 2000. *Between politics and science*. Cambridge: Cambridge University Press.

Gutmann, A., and D. Thompson. 1996. *Democracy and disagreement*. Cambridge, MA: Harvard University Press.

Haack, S. 2003. *Defending science within reason*. New York: Prometheus Books.

Hacking, I. 2001. *The social construction of what?* Cambridge, MA: Harvard University Press.

Hall, H., and R. Ziedonis. 2001. The patent paradox revisited: An empirical study of patenting in the U.S. semiconductor industry. *Rand Journal of Economics* 32:101–28.

Hamilton, R. 2000. *The law of corporations*. 5th ed. St. Paul, MN: West Publishing.

Harding, S. 1986. *The science question in feminism*. Ithaca, NY: Cornell University Press.

Harris, G. 2004. New York state official sues drug maker over test data. *New York Times* (June 3, 2004): A1.

Harvard University Medical School. 1999. Authorship guidelines. http://www.hms.harvard.edu/integrity/authorship.html (accessed May 12, 2006).

Haskins, C. 1957. *The rise of universities*. Ithica, NY: Cornell University Press.

Heller, M., and R. Eisenberg. 1998. Can patents deter innovation? The anticommons in biomedical research. *Science* 280:698–701.

Hilts, P. 1997. Researcher profited after study by investing in cold treatment. *The New York Times* (February 1, 1997): A6.

Holden, C., and G. Vogel. 2002. "Show us the cells," U.S. researchers say. *Science* 297:923–25.

Honderich, T. ed. 1995. *The Oxford companion to philosophy*. New York: Oxford University Press.

Hull, D. 1988. *Science as a process*. Chicago: University of Chicago Press.

Human Genome Project (HGP). 2003a. History of the Human Genome Project. http://www.ornl.gov/TechResources/Human_Genome/project/hgp.html (accessed May 12, 2006).

———. 2003b. Ethical, legal, and social issues. http://www.ornl.gov/TechResources/Human_Genome/elsi/elsi.html (accessed May 12, 2006).

Hume, D. 1748. *An enquiry concerning human understanding*. Indianapolis: Hackett, 1993.

Huth, E. 1986. Irresponsible authorship and wasteful publication. *Annals of Internal Medicine* 104:257–59.

———. 2000. Repetitive and divided publication. In *Ethical issues in biomedical publication*, ed. A. Jones and F. McLellan, 112–36. Baltimore, MD: Johns Hopkins University Press.

ICMJE (International Committee of Medical Journal Editors). 2005. Uniform requirements for manuscripts submitted to biomedical journals. http://www.icmje.org/index.html (accessed May 12, 2006).

Institute of Medicine. 2005. *Complementary and alternative medicine in the United States*. Washington, DC: National Academy Press.

International Committee of Medical Journal Editors. *See* ICMJE.

International Standard Randomized Controlled Trial Number. *See* ISRCTN.

Iribarne, A. 2003. Orphan diseases and adoptive initiatives. *Journal of the American Medical Association* 290:116.

ISRCTN (International Standard Randomized Controlled Trial Number). 2003. Why register trials? http://controlled-trials.com/isrctn/why_register.asp (accessed May 12, 2006).

Jaffe, A. 1996. Patterns and trends of research and development expenditures in the United States. *Proceedings of the National Academy of Sciences*. 93: 12658–63.

Jaffe, A., and J. Lerner. 2004. *Innovation and its discontents*. Princeton, NJ: Princeton University Press.

JAMA (*Journal of the American Medical Association*). 2003. Authorship requirements. http://jama.ama-assn.org/ifora_current.dtl (accessed May 12, 2006).

JAMA Editors. 1999. *JAMA* and editorial independence. *Journal of the American Medical Association* 281, 460.

James, W. 1898. *The will to believe*. New York: Dover, 1956.

Jones, A. 2000. Changing traditions of authorship. In *Ethical issues in biomedical publication*, ed. A. Jones and F. McLellan, 3–29. Baltimore, MD: Johns Hopkins University Press.

Journal of the American Medical Association. *See JAMA*.

Kaiser, J. 1999. Plan for divulging raw data eases fears. *Science* 283:914–15.

———. 2003. NIH roiled by inquiries over grants hit list. *Science* 302:758.

———. 2004a. Senators probe alleged financial conflicts at NIH. *Science* 303:603–604.

———. 2004b. Conflict of interest: Report suggests NIH weigh consulting ban. *Science* 305:1090.

———. 2004c. House votes to kill grants, limit travel to meetings. *Science* 305:1688.

Kant, I. 1753. *Grounding for the metaphysics of morals*. Trans. J. Ellington. Indianapolis: Hackett, 1981.

———. 1787. *Critique of pure reason*. Trans. N. Smith. New York: Macmillan, 1985.

Karp, J. 1991. Experimental use as patent infringement: The impropriety of broad exemption. *Yale Law Journal* 100:2169–88.

Kassirer, J. 1999. Editorial independence. *The New England Journal of Medicine* 340:1671–72.

———. 2004. Why should we swallow what these studies say? *The Washington Post* (August 1, 2004): B3.

Kealey, T. 1997. *The economic laws of scientific research*. New York: Macmillan.

Kerr, C. 2001. *The uses of the university*, 5th ed. Cambridge, MA: Harvard University Press.

Kim, S., R. Millard, P. Nisbet, C. Cox, and E. Caine. 2004. Potential research participants' views regarding researcher and institutional financial conflicts of interest. *Journal of Medical Ethics* 30:73–79.

Kintisch, E. 2005. Researcher faces prison for fraud in NIH grant applications and papers. *Science* 307:1851.

Kitcher, P. 1993. *The advancement of science*. New York: Oxford University Press.

———. 2001. *Science, truth, and democracy*. New York: Oxford University Press.

Klee, R. 1997. *Introduction to the philosophy of science*. New York: Oxford University Press.

Knorr-Cetina, K. 1981. *The manufacture of knowledge*. Oxford: Pergamon Press.

Krimsky, S. 2003. *Science in the private interest*. Lanham, MD: Rowman and Littlefield.

Krimsky, S., L. Rothenberg, P. Stott, and G. Kyle. 1996. Financial interests of authors in scientific journals: A pilot study of 14 publications. *Science and Engineering Ethics* 2:395–420.

Kuhn, T. 1970. *The structure of scientific revolutions*. 2nd ed. Chicago: University of Chicago Press. (1st ed. 1962.)

———. 1977. *The essential tension*. Chicago: University of Chicago Press.

LaFollette, M. 1992. *Stealing into print*. Berkeley: University of California Press.

Laudan, L. 1984. *Science and values*. Berkeley: University of California Press.

Latour, B., and S. Woolgar. 1986. *The social construction of facts*. Princeton, NJ: Princeton University Press.

Lawler, A. 2004. Harvard enters stem cell fray. *Science* 303:1453.

Leshner, A. 2003. Don't let ideology trump science. *Science* 302:1479.

Lexchin, J., L. Bero, B. Djulbegovic, and O. Clark. 2003. Pharmaceutical industry sponsorship and research outcome and quality: Systematic review. *British Medical Journal* 326:1167–70.

Lidz, C., and P. Appelbaum. 2002. The therapeutic misconception: Problems and solutions. *Medical Care* 40 (9 Suppl): V55–63.

Lo, B., L. Wolf, and A. Berkeley. 2000. Conflict of interest policies for investigators in clinical trials. *New England Journal of Medicine* 343: 1616–20.

Locke, J. 1690. *An essay concerning human understanding*. New York: Prometheus Books, 1994.

Lockwood, A. 2004. Human testing of pesticides: Ethical and scientific considerations. *American Journal of Public Health* 94:1908–16.

Longino, H. 1990. *Science as social knowledge*. Princeton, NJ: University of Princeton Press.

Lycan, W. 1988. *Judgment and justification*. Cambridge: Cambridge University Press.

MacIlwain, C. 1999. Scientists fight for right to withhold data. *Nature* 397:459.

Madey v. Duke University. 307 F.3d 1351 (2002).

Magnus, D., A. Caplan, and G. McGee, eds. 2002. *Who Owns Life?* Amherst, NY: Prometheus Books.

Malakoff, D. 2003. Opening the books on open access. *Science* 302:550–54.

———. 2004. 2005 budget makes flat a virtue. *Science* 303:748–50.

Marshall, E. 1999a. Two former grad students sue over alleged misuses of ideas. *Science* 284:562–63.

———. 1999b. Universities balk at OMB funding rules. *Science* 278:1007.

———. 2001a. Sharing the glory, not the credit. *Science* 291:1189–93.

———. 2001b. Testing time for missile defense. *Science* 293:1750–52.

———. 2001c. Gene therapy in trial. *Science* 288:951–54.

———. 2001d. Appeals court clears way for academic suits. *Science* 293: 411–12.

Martin, B. 1995. Against intellectual property. *Philosophy and Social Action* 21 (3): 7–22.

May, R. 1999. The scientific investment of nations. *Science* 281:49–55.

Mayo, D. 1996. *Error and the growth of experimental knowledge*. Chicago: University of Chicago Press.

Mayr, E. 1982. *The growth of biological thought*. Cambridge, MA: Harvard University Press.

McCrary, V., C. Anderson, J. Jakovljevic, T. Khan, L. McCullough, N. Wray, and B. Brody. 2000. A national survey of policies on disclosure of conflicts of interest in biomedical research. *New England Journal of Medicine* 343:1621–26.

McLellan, F. 2000. Ethics in cyberspace: The challenges of electronic scientific publication. In *Ethical issues in biomedical publication*, ed. A. Jones and F. McLellan, 166–95. Baltimore, MD: Johns Hopkins University Press.

McPherson, M., and M. Schapior. 2003. Funding roller coaster for public higher education. *Science* 302:1157.

Meadows, J. 1992. *The great scientists*. New York: Oxford University Press.

Merges, R., and R. Nelson. 1990. On the complex economics of patent scope. *Columbia Law Review* 90, 839–916.

Merton, R. 1938. Science and the social order. *Philosophy of Science* 5:321–37.

———. 1973. *The sociology of science*. Chicago: University of Chicago Press.

———. 2000. Letter to David B. Resnik, February 5, 2000.

Miller, A., and M. Davis. 2000. *Intellectual property*. St. Paul, MN: West Publishing.

Miller, T., and C. Horowitz. 2000. Disclosing doctors' incentives: Will consumers understand and value the information? *Health Affairs* 19:149–55.

Moore v. Regents of the University of California. 793 P.2d 479 (Cal. 1990).

Morin, K. 1998. The standard of disclosure in human subject experimentation. *Journal of Legal Medicine* 19:157–221.

Morin, K., H. Rakatansky, F. Riddick Jr, L. Morse, and J. O'Bannon III, M. Goldrich, P. Ray, M. Weiss, R. Sade, and M. Spillman. 2002. Managing conflict of interest in the conduct of clinical trials. *Journal of the American Medical Association* 287:78–84.

Moses, H., E. Dorsey, D. Matheson, and S. Their. 2005. Financial anatomy of biomedical research. *Journal of the American Medical Association* 294: 1333–42.

Moses, H., and B. Martin. 2001. Academic relationships with industry: A new model for research. *Journal of the American Medical Association* 285:933–35.

Murphy, D. 2004. Defying Bush administration, voters in California approve $3 billion for stem cell research. *New York Times* (November 5, 2004): A1.

Murphy, J. 2000. Expert witnesses at trial: Where are the ethics? *The Georgetown Journal of Legal Ethics* 14 (1): 217–239.

National Academy of Sciences (NAS). 1992. *Responsible science*. Washington, DC: National Academy Press.

———. 1995. *On being a scientist*. 2nd ed. http://www.nap.edu/readingroom/books/obas/ (accessed May 12, 2006).

———. 1997. *Advisor, teacher, role model, friend: On being a mentor to students in science and engineering*. Washington: NAS.

———. 2002. *Integrity in science*. Washington, DC: National Academy Press.

National Center for Biotechnology Information. *See* NCBI.

National Human Research Protections Advisory Committee. 2001. Draft interim guidance: Financial relationships in clinical research. http://www.hhs.gov/ohrp/nhrpac/mtg12-00/finguid.htm (accessed May 13, 2006).

National Institutes of Health. *See* NIH.

National Research Council. 1996. *Guide for the care and use of laboratory animals.* Washington, DC: National Academy Press.

National Science Foundation. *See* NSF.

NCBI (National Center for Biotechnology Information). 2003. GenBank Overview. http://www.ncbi.nlm.nih.gov/Genbank/ (accessed May 12, 2006).

NEJM (*New England Journal of Medicine*). 2003. Information for authors. http://www.nejm.org/general/text/InfoAuth.htm#Conflict (accessed May 13, 2006).

Newton-Smith, W. 1981. *The rationality of science.* New York: Routledge.

NIH. 1992. Guidelines for the conduct of research in the intramural programs at the NIH. Available at http://www.nih.gov/news/irnews/guidelines.htm#anchor128256 (accessed May 13, 2006).

———. 2002. Statement on sharing research data. http://grants1.nih.gov/grants/guide/notice-files/NOT-OD-02-035.html (accessed May 13, 2006).

———. 2005. Conflict of interest information and resources. http://www.nih.gov/about/ethics_COI.htm (accessed May 13, 2006).

North Carolina Biotechnology Center. 2004. About us. http://www.ncbiotech.org/aboutus/aboutus.cfm (accessed May 13, 2006).

NSF (National Science Foundation). 2002. *Science and engineering indicators.* http://www.nsf.gov/statistics/seind02/ (accessed May 13, 2006).

Odlyzko, A. 1995. Tragic loss or good riddance? The impending demise of traditional scholarly journals. In *Electronic publishing confronts academia: The agenda for the year* 2000, ed. R. Peek and G. Newby, 71–122. Cambridge, MA: MIT Press.

Office of Research Integrity (ORI). 1998. Scientific Misconduct Investigations, 1993–1997. http://ori.dhhs.gov/documents/misconduct_investigations_1993_1997.pdf (accessed May 13, 2006).

Office of Science and Technology Policy. 2000. Federal Research Misconduct Policy. *Federal Register* 65 (235): 76260–64.

Olivieri, N. 2003. Patients' health or company profits? The commercialization of academic research. *Science and Engineering Ethics* 9:29–41.

Olson, C., D. Rennie, D. Cook, K. Dickersin, A. Flanagin, J. Hogan, Q. Zhu, J. Reiling, and B. Pace. 2002. Publication bias in editorial decision making. *Journal of the American Medical Association* 287:2825–28.

Patton, P. 1995. Exposing the black budget. *Wired Magazine* (November 1995). http://wired.com/wired/archive/3.11/patton.html (accessed May 13, 2006).

Peirce, C. 1940. *Philosophical writings.* New York: Dover, 1955.

Pennock, R. 1996. Inappropriate authorship in collaborative scientific research. *Public Affairs Quarterly* 10:379–93.

Pharmaceutical Research and Manufacturers of America (PhRMA). 2003. Research and development. http://www.phrma.org/index.php?option=

com_content&task=view&id=123&Itemid=109&cat=Research+and+Development (accessed May 13, 2006).

PHS. *See* Public Health Service.

Pickering, A. 1992. *Science as practice and culture*. Chicago: University of Chicago Press.

Pollack, A. 2002. Genome pioneer will start center of his own. *The New York Times* (August 16, 2002): C1.

Popper, K. 1959. *The logic of scientific discovery*. New York: Routledge.

President's Council on Bioethics. 2002. *Human cloning and human dignity: An ethical inquiry*. http://www.bioethics.gov/reports/cloningreport/index.html (accessed May 13, 2006).

Press, E., and J. Washburn. 2000. The kept university. *The Atlantic Monthly. v285 n3 p39–42,44–52,54 Mar* 2000. http://www.theatlantic.com/issues/2000/03/press.htm (accessed May 7, 2006).

Public Health Service (PHS). 1995. Objectivity in research. *Federal Register* 60, 132: 35810–19.

Pulley, J. 2002. U. of Arkansas receives $300-million pledge, the largest ever to a public college. *Chronicle of Higher Education* 48 (33): A32.

Quine, W. 1961. *Word and object*. Cambridge. MA: MIT Press.

———. 1977. *Ontological relativity and other essays*. New York: Columbia University Press.

———. 1986. *Theories and things*. Cambridge, MA: Harvard University Press.

Quine, W., and J. Ullian. 1978. *The web of belief*. 2nd ed. New York: Random House.

Rawls, J. 1993. *Political liberalism*. New York: Columbia University Press.

Relman, A. 1999. The NIH "E-biomed" proposal: A potential threat to the evaluation and orderly dissemination of new clinical studies. *New England Journal of Medicine* 340:1828–29.

Rennie, D. 1999. Fair conduct and fair reporting of clinical trials. *Journal of the American Medical Association* 282:1766–68.

———. 2000. Improving the conduct and reporting of clinical trials. *Journal of the American Medical Association* 283:2787–90.

Rennie, D., V. Yank, and L. Emanuel. 1997. When authorship fails: A proposal to make authors more accountable. *Journal of the American Medical Association* 278:579–85.

Resnik, D. 1992a. Are methodological rules hypothetical imperatives? *Philosophy of Science* 59:498–507.

———. 1992b. The fittingness theory of truth. *Philosophical Studies* 68:95–101.

———. 1993. Do scientific aims justify methodological rules? *Erkenntnis* 39: 223–32.

———. 1996. Data falsification in clinical trials. *Science Communication* 18 (1): 49–58.

———. 1997. A proposal for a new system of credit allocation in science. *Science and Engineering Ethics* 3:237–44.

———. 1998a. *The ethics of science*. New York: Routledge.

———. 1998b. The ethics of HIV research in developing nations. *Bioethics* 12:285–306.

———. 1998c. Conflicts of interest in science. *Perspectives on Science* 6:381–408.

———. 2000*a*. *Financial interests and research bias*. Perspectives on Science 8*:* 255–85.

———. 2000*b*. *Statistics, ethics, and research: An agenda for education and reform.* Accountability in Research 8*:*163–88.

———. 2001a. Setting biomedical research priorities: Justice, science, and public participation. *Kennedy Institute for Ethics Journal* 11:181–205.

———. 2001b. DNA patents and scientific discovery and innovation: Assessing benefits and risks. *Science and Engineering Ethics* 7:29–62.

———. 2001c. Developing drugs for the developing world: An economic, legal, moral, and political dilemma. *Developing World Bioethics* 1:11–32.

———. 2003a. Are DNA patents bad for medicine? *Health Policy* 65:181–97.

———. 2003b. From Baltimore to Bell Labs: Reflections on two decades of debate about scientific misconduct. *Accountability in Research* 10:123–35.

———. 2003c. A pluralistic account of intellectual property. *Journal of Business Ethics* 46:319–35.

———. 2003d. Patents and the research exemption. *Science* 299:821–22.

———. 2003e. Setting biomedical research priorities in the twenty-first century. *Virtual Mentor* 5:7. http://www.ama-assn.org/ama/pub/category/10571.html (accessed May 14, 2006).

———. 2003f. Strengthening the United States' database protection laws: Balancing public access and private control. *Science and Engineering Ethics* 9:301–18.

———. 2003g. A biotechnology patent pool: An idea whose time has come? *The Journal of Philosophy, Science, and Law* 3 (January 2003). http://www.miami.edu/ethics/jpsl/archives/papers/biotechPatent.html (accessed May 14, 2006).

———. 2004a. The distribution of biomedical research resources and international justice. *Developing World Bioethics* 4:42–57.

———. 2004b. Disclosing financial interests to research subjects: Ethical and legal Issues. *Accountability in Research* 11:141–59.

———. 2004c. *Owning the genome.* Albany, NY: SUNY Press.

———. 2004d. Liability for institutional review boards: From regulation to litigation. *Journal of Legal Medicine* 25:131–84.

Resnik, D., and A. Shamoo. 2002. Conflict of interest and the university. *Accountability in Research* 9:45–64.

Roberts, L. 2001. Controversial from the start. *Science* 291:1182–88.

Rochon, P., J. Gurwitz, R. Simms, P. Fortin, D. Felson, K. Minaker, and T. Chalmers. 1994. A study of manufacturer-supported trials of non-steroidal anti-inflammatory drugs in the treatment of arthritis. *Archives of Internal Medicine* 154:157–63.

Rosenberg, A. 1988. *Philosophy of social science.* Boulder, CO: Westview Press.

———. 2000. *Philosophy of science.* New York: Routledge.

Rowen, L., G. Wong, R. Lane, and L. Hood. 2001. Publication rights in the era of open data release policies. *Science* 289:1881–82.

Salary.com. 2003. www.salary.com (accessed May 14, 2006).

Sanders, S., and J. Reinisch. 1999. What would you say if...? *Journal of the American Medical Association* 281: 275–77.

Schaffner, K. 1986. Ethical problems in clinical trials. *Journal of Medicine and Philosophy* 11:297–315.

Scheffler, I. 1967. *Science and subjectivity.* Indianapolis: Bobbs-Merrill.

Schwartz, L. 2002. Media coverage of scientific meetings: Too much, too soon? *Journal of the American Medical Association* 287:2859–63.

Science. 2003. Patent factories. *Science* 299:1511.

Service, R. 2000. Can Celera do it again? *Science* 287:2136–38.

Shamoo, A., and D. Resnik. 2003. *Responsible conduct of research.* New York: Oxford University Press.

Shimm, D., R. Spece, M. DiGregorio. 1996. Conflicts of interest in relationships between physicians and the pharmaceutical industry. In *Conflicts of Interest in Clinical Practice and Research,* ed. D. Shimm, R. Spece, and A. Buchanan, 321–60. New York: Oxford University Press.

Schrader-Frechette, K. 1994. *The ethics of scientific research.* Lanham, MD: Rowman and Littlefield.

Simes, R. 1986. Publication bias: The case for an international registry of clinical trials. *Journal of Clinical Oncology* 4:1529–41.

Smith, K. 1997. CU, profs to get $45 million. *The Denver Post* (July 8, 1997): A1.

Solomon, M. 2001. *Social empiricism.* Cambridge, MA: MIT Press.

Stelfox, H., G. Chua, K. O'Rourke, and A. Detsky. 1998. Conflict of interest in the debate over calcium channel antagonists. *New England Journal of Medicine* 338:101–6.

Steneck, N. 2000. Assessing the integrity of publicly funded research. In *Proceedings for the Office of Research Integrity's conference on research on research integrity,* 1–16. Washington, DC: Office of Research Integrity.

———. 2004. *ORI introduction to responsible conduct of research.* Washington: Office of Research Integrity.

Strom, S. 2002. Universities report record in private contributions. *New York Times* (March 22, 2002): A14.

Studdert, D., M. Mello, and T. Brennan. 2004. Financial conflicts of interest in physicians' relationships with the pharmaceutical industry—Self-regulation in the shadow of federal prosecution. *New England Journal of Medicine* 351:1891–900.

Svatos, M. 1996. Biotechnology and the utilitarian argument for patents. *Social Philosophy and Policy* 13:113–44.

Swazey, J., M. Anderson, and K. Lewis. 1993. Ethical problems in academic research. *American Scientist* 81:542–53.

Tauer, C. 2002. Central ethical dilemmas in research involving children. *Accountability in Research* 9:127–42.

Taylor, R., and J. Giles. 2005. Cash interests taint drug advice. *Nature* 437: 1070–71.

Teitelman, R. 1994. *The profits of science*. New York: Basic Books.

Thagard, P. 1992. *Conceptual revolutions*. Princeton, NJ: Princeton University Press.

Thompson, D. 1993. Understanding financial conflicts of interest. *New England Journal of Medicine* 329:573–76.

Thomsen, M., and D. Resnik. 1995. The effectiveness of the erratum in avoiding error propagation in physics. *Science and Engineering Ethics* 1:231–40.

Thomson, J., J. Itskovitz-Eldor, S. Shapiro, M. Waknitz, J. Swiergiel, V. Marshall, and J. Jones. 1998. Embryonic stem cell lines derived from human blastocysts. *Science* 282:1145–47.

Thursby, J., and M. Thursby. 2003. University licensing and the Bayh-Dole Act. *Science* 301:1052.

University of Louisville. 2002. Faculty salary analysis. http://institutionalresearch.louisville.edu/files/ir/facsalary/fs200001.pdf (accessed May 13, 2006).

U.S. Census Bureau. 2003. Median income for four-person families, by state. http://www.census.gov/hhes/income/4person.html (accessed May 14, 2006).

U.S. Congress, Committee on Government Operations. 1990. Are scientific misconduct and conflicts of interest hazardous to our health? Report 101–688. Washington: U.S. Government Printing Office.

U.S. Constitution. 1787. Article 1, Section 8, Clause 8.

U.S. Patent Act. 1995. 35 United States Code 101.

U.S. Patent and Trademark Office (USPTO). 2003. Glossary. http://www.uspto.gov/main/glossary/index.html#cfr (accessed May 14, 2006).

Van Fraassen, B. 1980. *The scientific image*. Oxford: Clarendon Press.

Wadman M. 2004. NIH head stands firm over plans for open access. *Nature* 432:424.

Webster's Ninth New Collegiate Dictionary. 1983. Springfield, MA: Merriam-Webster Inc.

Weiss, R. 2005. Bioethics council head to step down. *Washington Post* (September 9, 2005): A6.

Weissman, P. 2002. Stem cells—Scientific, medical, and political issues. *New England Journal of Medicine* 346:1576–79.

Whittington, C., T. Kendall, P. Fonagy, D. Cottrell, A. Cotgrove, and E. Boddington. 2004. Selective serotonin reuptake inhibitors in childhood depression: Systematic review of published versus unpublished data. *The Lancet* 363:1341–45.

Wible, J. 1998. *The economics of science*. New York: Routledge.

Wilcox, L. 1998. Authorship: The coin of the realm, the source of complaints. *Journal of the American Medical Association* 280:216–17.

Williams, T. 1987. *The history of invention.* New York: Facts on File Publications.

Willman, D. 2003. Stealth merger: Drug companies and government medical research. *Los Angeles Times* (December 7, 2003): A1.

Wittgenstein, L. 1953. *Philosophical investigations.* Trans. G. Anscombe. New York: MacMillan.

Wolfe, S. 2003. Interview with Sidney Wolfe. Frontline. Public Broadcasting System. http://www.pbs.org/wgbh/pages/frontline/shows/prescription/interviews/wolfe.html (accessed May 14, 2006).

WordNet. 1997. http://dictionary.com (accessed May 14, 2006).

Ziman, J. 1968. *Public knowledge.* Cambridge: Cambridge University Press.

———. 2000. *Real science.* Cambridge: Cambridge University Press.

INDEX